THE EXPLODING POPULATION BOMB:

SOCIETIES UNDER STRESS
CORRECTIVE STRATEGIES AND SOLUTIONS

FREDERIC R. SIEGEL PH.D.
PROFESSOR EMERITUS
GEORGE WASHINGTON UNIVERSITY

STRESSORS
WATER AND FOOD INSECURITY
SOCIO-ECONOMIC UNCERTAINTY
POLITICAL IMPOTENCE/INDIFFERENCE
NATURAL HAZARDS - GLOBAL WARMING/CLIMATE CHANGE

DEDICATION

I dedicate this book to those who struggle today to confront socio-political, economic, and environmental problems that re-press societal development for billions of people worldwide. I dedicate this book to future generations who will inherit these and other problems during coming decades and work to face them down. Omni-national efforts can find success in these endeavors if governments work together without the specter of national goals or pride but with the purpose of solving problems that affect all of humankind.

Frederic R. Siegel
Washington, D.C.
March 2010

PROLOGUE

Rachel Carson published the book "Silent Spring" in 1962 and gave great impetus to our modern environmental thought and the ensuing environmental movements. In the book she emphasized the harmful effects of pesticides on animals and humans, focusing on DDT and birds.[1] Paul R. Ehrlich, surely inspired by "Silent Spring", published the book "The Population Bomb" in 1968. In it he espoused that populations were growing at a rate that was faster than the rate at which we could produce food and provide other natural resources. If the fertility rate was not brought down to the replacement level, the population bomb would explode and 100s of millions would die from starvation and disease.[2]

When Ehrlich wrote "The Population Bomb", the global population was 3.5 billion people. Twenty-two years later (1990), he and co-author Anne H. Ehrlich published the book "The Population Explosion."[3] By that time the world population had grown to 5.3 billion people but the "Green Revolution" and good food production practices avoided the predicted famine, but malnutrition (under-nourishment) was a problem for hundreds of millions, as it is today. By the end of 2010, 42 years later, the number of people on Earth will be close to double that in 1968, or almost 7.0 billion citizens.

In truth, there has been a population explosion, mainly in less developed and developing countries in Africa and Asia. It is not only a lack of safe water and untainted nutritious food that threatens increasing populations in these countries. Several other natural and anthropogenic factors put today's populations and future generations at risk. These include global warming/ climate change, pollution, competition among people and between nations for natural resources that fuel economic development, and disease and natural disasters that will impact our living environments. "Act now or suffer the consequences" can be taken as an important thesis of the forenamed books. This book brings forward the discussion of population growth. It examines the stresses on people and governments from the demands of expanding populations on ecosystems that sustain them. The book further evaluates the potential that exists to ameliorate natural and anthropogenic problems that assault the well-being of people and their quality of life

Commodity traders are arch-typical speculators. They bet on the future when buying commodities at a point in time with the expectation that they will sell them for higher prices at a later time. This is often not the case and traders (i.e., investors) end up selling the futures at a loss. As will be emphasized in this book, such speculation is not the norm for demographers who project population figures for the next fifty or more years. Population figures are as close to sure things as you can get.

Demographers predict that by 2050 the expanded world population will reach 9.5 billion people or more, or about 40% higher than today's 6.8 billion inhabitants of Planet Earth. This global growth in population is not spread evenly among all geographic regions and the nations that they encompass. It is most pronounced in Africa, parts of Asia, the Middle East, and South America. Within Africa, for example, population growth is highest in the sub-Saharan zone, especially in Nigeria, the Congo (Kinshasa) and Uganda.

Indeed, given the number of persons in these regions under 15 years old and coming into childbearing age, and those in early to middle childbearing years (15-35/40), it is clear that the world is flirting with a real problem...an intensifying population expansion that could evolve into a threat to societal stability in nations least able to cope with it. This could be a global catastrophe and likely lead to collapses of some societies and war among others. If abstention is not followed, expedient and widespread focused educational programs that incorporate the importance of the use of contraceptive methods can be effective in two ways. First, the programs begin to inculcate the notion of protection against sexually transmitted diseases. Second, they can prevent unwanted pregnancies and help to limit population growth. Conversely, populations are contracting in Japan and several Western and Eastern European countries creating unique problems for them. Nonetheless, these countries will also feel the impacts of a greater number of inhabitants on the Earth both from immigration (legal and illegal) and from accessibility to and competition for natural resources.

Whether we consider expanding or contracting populations, stresses that affect people's lives as to physical health, emotionally, socially, and economically, build up and are a cause of concern for the well being of societies. International financial institutions (e.g., the World Bank, regional development banks), national aid agencies, and NGOs must work to mitigate the many and varied pressures on a nation's citizenry. The degree to which they do this will determine whether or not there can be stable, healthy, and thriving societies.

We will evaluate various technological, social, and economic approaches that can be used to devise solutions that can ease the problems now experienced by expanding populations and also by aging and contracting ones. We will review the positive lessons learned in mitigation or solution of past and existing problems related to servicing societal needs. This will provide guidance on how to rework some of the lessons so that they can be used to alleviate threats from population pressure points. Most critically, we will determine how these and new concepts can be applied to assuage tensions that will be experienced in the future by growing populations. An example is the stress from predicted climate changes caused by global warming, a process originated from human activities and abetted by them.

Clearly, we must plan to cope with the need to provide the basic requirements of life for human masses perhaps half again as large as at present. Especially important is access to water from sources that are irregularly distributed geographically. As important is access to agricultural commodities limited by cropland in production and by difficulties in food distribution. Populations must also be able to obtain many other natural resources that are needed for industrial and manufacturing development, and hence job creation (e.g., energy sources, fertilizer, metal ores, industrial minerals, wood, fibers). Nations must plan to provide the healthcare, educational opportunities, and gainful employment for an increased number of citizens. We must strive to create acceptance bridges to manage the needs of nations with aging and contracting populations where acceptance of foreign workers is difficult. This problem arises because of marked and sometimes divisive cultural or religious differences and by workers who cannot adapt to the values, customs, and traditions of a host country.

Problems have solutions, some easier to put in place than others. Putting just one solution in place is a start and could stimulate critical thinking and creative experimentation. Ultimately, this will increase the solutions inventory needed to face the stresses that develop as populations grow, and to some degree where they contract.

Mao Zedong wrote "The people, and the people alone, can be the motive force in the making of world history." Following this thought, "The people, and the people alone, can be the motive force in regulating population expansion, conserving natural resources, and reducing the stresses on themselves and on living environments that support their societies."

CONTENTS

LIST OF TABLES

LIST OF FIGURES

CHAPTER I. A GROWING GLOBAL POPULATION - WHEN WILL IT STOP?

GROWTH OF POPULATION
1900 to the Present (2010)

In 1900, the world population was estimated at 1.65 billion people. Sixty years later the population nearly doubled to 3.02 billion persons. By the year 2000, the population doubled to 6.06 billion people. In 2008 the Earth's population reached 6.68 billion inhabitants. Each decade through 1990, the average number of people added to the global population increased at a progressively higher rate with two exceptions: during the WWI decade (1910-1920) and during the WWII decade (1940-1950). This is illustrated in Table 1-1. From 1990 to 2008, the average annual growth in global population has lessened.

During WWII, the estimated death toll from the conflict was 60-72 million souls. If there had been no war, an average population growth rate of 1.4% for the decade 1940-1950 would have given an annual average growth rate estimated at 36 million people instead of the 22 million reported in Table 1-1. As a result of WWII, today's population (6.68 billion in 2008) is significantly less than it would have been without the war deaths, perhaps 7+ billion people. The global problems we have today with respect to lack of safe water for billions, poor nutrition and starvation for 100s of millions, worldwide pollution, and conflicts among and between populations would have been exacerbated.

Table 1-1. Historical Growth of Population (extracted from United Nations estimates).
http://www.un.org/esa/population/publications/sixbillion/six-bilpart1.pdf

	World Population	Total Growth During Decade (rounded)	Average Annual Growth During Decade
1900	1.65 billion		
1910	1.75	100 million	10 million
1920	1.86	110	11
1930	2.07	210	21
1940	2.30	230	23
1950	2.52	220	22
1960	3.02	500	50
1970	3.70	680	68
1980	4.44	740	74
1990	5.27	830	83
2000	6.06	790	79
2009	6.74	780	78

In the following two major sections, the data for fertility rates, natural rates of population growth, and populations in nations and regions are excerpted from or extrapolated from the 2009 World Population Data Sheet.[4]

FACTORS THAT INFLUENCE POPULATION GROWTH

Fertility Rate

A woman's childbearing years are from 15-49 years old. Early to middle childbearing ages of 15-35/40 are most prolific. The average number of children borne by women during these years is called the fertility rate. At a global fertility rate of 2.06, we have replacement of population without continuing growth. The fertility rate for the world in 2008 was 2.6. In the less developed/developing world excluding China the figure was 3.1 (with China and its one child per couple law, the rate was 2.7). In the developed world, the fertility rate was 1.7. The global fertility rate has been slowing since about 1950 (Table 1-2).

Demographers calculate that if fertility continues to slow at the same pace, the replacement rate will be reached by 2047. Until then, global population will continue to grow. The developed world with a fertility rate of 1.7 is well below the replacement rate (2.06) and as a result its population is contracting. It is clear that fertility rates are decreasing globally (Table 1-2), but the fertility rates for geographic regions between 1980 and 1999 vary greatly (Table 1-3). Table 1-3 shows that the Sub-Saharan Africa region, the Middle East and North Africa, and South Asia are at risk of unsustainable population expansion. These regions must lower their fertility rates to supportable levels in the near future, at the least within two generations.

Table 1-2. Average number of children per woman, worldwide, 1950-2050.[5]

Time Frame	1950 1955	1970 1975	1990 1995	2010 2015	2030 2040	2040 2050
Fertility Rate	5.0	4.5	2.9	2.4	2.1	2.0

It is possible to bring down fertility rates markedly in a two generations time frame. For example, from 1950-1955 to 2008, the fertility rate in Bangladesh decreased from 6.7 to 2.5. Similarly, the rates dropped in Mexico from 6.7 to 2.3, and in South Africa from 6.5 to 2.7. However, and in contrast and to the detriment of their countries, the fertility rate during the same period in Uganda went from 6.9 to only 6.7, and the rates in Niger and Yemen dropped from 8.1 to 7.4, and from 8.2 to 5.5, respectively.

Table 1-3. Geographic regions and their average fertility rates, births per woman, 1980 to 1999 (modified).[6]

Region	1980	1999
Low and middle income countries	4.1	2.9
Sub-Saharan Africa	6.6	5.3
Middle East and North Africa	6.1	3.5
South Asia	5.3	3.4
Latin America and Caribbean	4.1	2.6
East Asia and Pacific	3.0	2.1
Eastern Europe and Central Asia	2.5	1.6
High Income Countries	2.5	1.7

At a fertility rate that shrinks to 1.96 (below replacement level), the world population would ultimately stabilize at about 6.5 billion people, the 2007 number. The stabilization figure would be about 11.5 billion people if the rate stayed at 2.16 (above replacement level). When the fertility rate fixes at 2.06, the global population is projected to settle at 9.5 billion persons. In either of the latter two cases, the question is, can we sustain these populations almost half or more than the population in 2008 when we did not, in 2008, achieve an acceptable quality of life for billions of world citizens?

Causes of High Fertility

There are many factors that result in high fertility rates. One lies in religious beliefs with respect to birth control methods other than abstinence. A second is that in some societies there is a tradition of having enough children to support parents when they can no longer work. Another is a lack of family planning education for females (and males) as to the benefits for their children when there are fewer mouths to feed and intellect to nourish. As these factors are modified in light of the global problems brought on by population growth, all peoples will ultimately benefit. Lastly, but as important if not more so than the factors given above, empowerment of women to learn, earn, vote, and influence government policies certainly has had a leading role in bringing down fertility rates in enlightened nations.

Rate of Population Growth

Doubling Time

The time it takes for a population to double is called the doubling time. Doubling time is calculated by dividing the number 70 by the percent rate of population growth. The percent rate of population growth is calculated from the birth rate per 1000 population minus the death rate per 1000 population. This is modified somewhat by emigration from a country and immigration into a country.

Thus, Saudi Arabia with a 2.6% growth rate (fertility rate 3.9) plans to increase its population in 27 years (~2035) from 28.7 million Saudis (+5.6 million guest workers) to 45.1 million Saudis. This presents a challenge to a country that services 20% of its population with potable water from high operational cost desalinization plants. Can the water supply be maintained when income from lower oil prices (October 2008 high $147 a barrel to November 2008 $51 a barrel to July 2009 $60 a barrel) and/or petroleum reserves deplete? The Congo (Kinshasa) with a population of 66.5 million people and a growth rate of 3.1% (fertility rate 6.5) could double the population to 133 million in 22 years. Similarly, Nigeria with a growth rate of 2.6% (fertility rate 5.7) could see its population rise from 148.1 million to 296.2 million citizens in 27 years. Given the political and social problems in these latter two countries now, what can be expected with twice their 2008 populations? Many other countries in Africa, Asia, and the Middle East are at risk from doubling their population in relatively short terms thus presenting problems of sustaining reasonable qualities of life for their people.

Social, economic, and political stresses in many countries are further heightened when we realize that more than 40% of their populations are under 15 years of age and will come into childbearing years. For Nigeria the figure is 45% and for the Congo (Kinshasa) it is 47%. Uganda, with a population of 29.2 million, has the largest population less than 15 years of age at 49%. At a growth rate of 2.9% (fertility rate 6.7) Uganda may double its population to 58.4 million citizens in 24 years. These are countries among many others with citizenries now suffering from frustration, hopelessness, and desperation. Such conditions will intensify dangerously with increasing populations and likely lead to violent civil unrest.

The growth in young childbearing populations (15-24 years of age) between 1950 and 2050 is projected to be the greatest in actual numbers in the Asia/Pacific region and Africa as shown in Table 1-4. As a percent of population, Africa shows the highest gain but the Asia/Pacific region has the highest percent of population in the age range. The Latin America/Caribbean region has a significant population gain but the percent of population 15-24 years of age remains stable. Conversely, more developed countries have a slight drop in population numbers but their percent of population in the 15-24 years old range drops markedly.

Table 1-4. The world's youth population, ages 15 to 24, 1950 and projected to 2050. Note the dramatic increases for Africa, Asia/Pacific, and Latin America/Caribbean regions (adapted from a figure in United Nations Population Division, 2009).[7]

Region	1950		2050	
	Population	%	Population	%
Africa	43 million	9	348 million	29
Asia/Pacific	248 million	54	639 million	53
Latin America/Caribbean	32 million	7	97 million	7
More Developed Countries	138 million	30	134 million	11

Conversely, Japan and several countries in Europe have zero or negative growth rates and contracting populations. Japan with a -0.0% growth rate (fertility rate 1.4) has 127.6 million people with 23% of its population over 65 years old and 13% under 15.

Its population is expected to shrink to 95.2 million people by 2050. This presents its own problems in terms of people to service the country's needs. Russia has a -0.3% growth rate (fertility rate 1.5) and its population is projected to shrink from 141.9 million people to 116.9 million by 2050. Germany and Ukraine have negative percent growth rates and as with Russia are projected to have large contractions in populations and the social, economic, and political problems that ensue from them.

POTENTIAL FOR POPULATION CRASH/DISRUPTION/STRIFE

The Earth is the master ecosystem for human society and contains subject ecosystems that support life. Populations can crash when the biological needs to sustain its life forms are no longer available in an ecosystem. This can cause massive die offs with only the strongest life forms surviving to reestablish societies that can be supported under prevailing conditions.

We can examine this concept of population crash using an example of guppies with a doubling time of one month in a fish tank that can sustain 1000 guppies. Two guppies are the initial inhabitants of the fish tank ecosystem. After one month there are four guppies whose number double sequentially to 8, 16, 32, 64, 128, 256, and 512 guppies during the following 7 months. The ecosystem sustains this population without problems, but toward the end of the 9th month, the guppy population exceeds the carrying capacity of the fish tank. The population crashes with most of the guppies dying out. The strong ones that remain then strive to reestablish a sustainable population.

Can this happen to the human population in the master ecosystem? To a degree, yes, but in not so drastic a crash mode but one that can be expected to severely impact the less- advantaged segments of national and regional societies. The result will be that survivors will work with advantaged societies to resolve the problems that brought on the crash. Once this is underway, the survivors will strive to bring populations back to numbers that their ecosystems can accommodate.

Realistically, populations can be disrupted in other ways that can drain the very essence of life from a society. There are the economic, social, and political ills cited previously. Although these ills will not kill people, they can wreak havoc in societies and drain their hopes and aspirations. This does long term damage to societal psyches and rends their very fabric.

Regional

The regions most at risk for increasing social disruption and physical strife are in Western, Eastern, and Middle Africa. Thirty-seven of the 44 countries that comprise these regions have doubling times of 35 or less years. Of these, 9 have growth rates of 3% or higher and 28 have rates of 2%-2.9% or higher. This is out of the 53 countries in Africa. On the average, Western Africa with a population of 297 million people has a growth rate of 2.7% (fertility rate 5.5). A doubling time of 26 years, modified by gradually falling fertility and growth rates, predicts that the 2035 population will be 501 million people. Similarly, Eastern Africa with a 2.6% growth rate (fertility rate 5.4) and a population of 313 million is projected to reach 545 million by 2035. The 2.8% growth rate (fertility rate 6.1) for Middle Africa will increase the region's population from 125 million to 236 million people by 2035. Much of the populace in these regions, with a 735 million people total in 2008 suffer from malnutrition, lack of safe water, limited health care possibilities for a great majority of their inhabitants, lack of educational opportunities, and high unemployment. This is a sure formula for social disorder as a cry for change to improve the societal quality of life.

Many nations have internal conflicts that result from political instability, lack of confidence in leadership, and poor security (vis a vis Congo [Kinshasa], Sudan, Somalia). The situation in Western, Eastern, and Middle Africa with 1.11 billion people in 2035 puts fear in the hearts and minds of governments that will have to deal with a regional population that is 51% larger than that in 2008. The risk of regional crashes because of the lack of basic needs to sustain life increases as social, economic, and political stability diminishes. Only immediate intervention by international institutions (e.g., the World Bank and regional development banks), advantaged nations (various Agencies for Development), and NGOs can drive the "now" planning and implementation of programs that will help reduce the toll from the sure to come crashes.

The Middle East with its 10 Arab nations (excluding Turkey), the Palestinian Territories, and Israel is a region at risk as the population grows at an annual rate of greater than 2% (fertility rate 3.3) from 138 million people in 2008 to 215 million in 2035. This may be underestimated because 34% of these populations are less than 15 years of age and coming into childbearing years in societies with high fertility rates. The rates range from 2.0 in Bahrain and the UAE to 4.6 in the Palestinian Territories to 5.5 in Yemen. An average of only 3.4% of the population is over 65 years old. War between nations over territory and water rights increases the regional risk of political crashes. The risks are compounded by conflicts between religious groups (Islamic Fundamentalists against Jewish and Christian believers), and internecine conflicts between groups competing for political power (e.g., Hamas and the Palestinian Authority, Hezbollah within the Lebanese government, Shia Islam and Sunni Islam in Iraq). In many cases, the Middle East regional situation risk condition is further compounded by government repression of dissenting political groups and limited acceptance of human rights in several nations.

Economic strife may occur in Eastern Europe as the result of a growth rate of -0.2% (fertility rate 1.5) and a population that will decrease from 295 million people in 2008 to 264 million in 2035. This will decrease governments' tax bases and limit their ability to deliver social services to segments of the population that need them most such as the aging and the very young. With fewer people to work on development projects, investment in a country will decline. The loss of an educated cadre to countries with better economic possibilities works against attracting investment. These factors, plus an increasingly older population that draws heavily on government resources to provide its needs and services, make economic progress questionable and economic downturn a possibility.

Malnutrition in young children, whether in low-income countries or income-advantaged nations, is of special concern and deserves discussion. Malnutrition can cause stunting usually during a child's first two years when they need a sufficient intake of nutrients from enough foods of good quality to grow normally. Stunting diminishes a child's cognitive abilities, limits educational potential, and hence future productivity as an adult. This can retard a nation's economic development if the problem is widespread as it is in many low-income countries. More than 40% of children under age 5 are stunted in 40 low-income countries, especially in Eastern Africa (50% stunted) and Middle Africa (40% stunted). India has 61 million stunted children under age 5. Although this is about 0.5% of the Indian population, it should be a serious concern for the country because as the Indian population grows so will the problem of malnutrition driven stunting. Of all of the countries in the world, only 21 have populations greater than 61 million people. Similarly, 41% of children under age 5 in South-Central Asia suffer from stunting.[8] Clearly, for the countries affected by the problem and for international agencies, malnutrition has to be a top priority problem to be resolved, first for socio-humanistic reasons and second for reaching economic development goals.

National

Japan has a zero growth rate and its population is projected to fall from 127.6 million people in 2008 to 109 million in

2035. As cited earlier, the country has the highest global percentage (23%) of citizens over 65 years old. Without enough workers to maintain its industrial base and agricultural production, and to care for its growing aged population in the near future, Japan may suffer a politico-economic crash unless it abrogates its policy to maintain cultural purity. In the very recent past, Japan has not welcomed guest workers to become citizens fully invested in its society. However, citizens of other countries (e.g., Brazil) who are of Japanese parentage have been sought to return to the home of their ancestors. Realism of national need should overcome the policy to maintain cultural purity.

Global

Global crashes have occurred in the past from great losses of population, causing damaging economic disruptions and socio-political upheavals. Nonetheless, human populations have survived and prospered. Not so in the case of animal kingdom population extinctions such as with dinosaurs and other life forms that could not cope with ecosystem changes or natural hazards. Human reasoning and logical thought, perseverance, ingenuity, and creativity have set the foundations for human populations to withstand terrible natural and anthropogenic-originated hazards and disasters.

MAJOR POPULATION LOSSES OTHER THAN FROM WAR

Disease

In the middle ages, the rat-borne oriental flea caused the black (bubonic) plague that ravaged China (and other areas of Asia) from 1318 to 1338. The plague brought the Chinese population down from 125 million to 90 million people, a 28% loss. Italian trading vessels carried the rat-borne plague to Europe in 1347. From 1347 to 1352 the disease reduced the population of Europe by a third, from 75 million to 50 million people. World population is estimated to have fallen from 450 million to 375 million inhabitants, an almost 17% global population loss. Humanity survived the plague-caused population crash by quarantine (isolation) of those with the disease from the unaffected populace. This was probably abetted by regulating sanitation conditions that sustained the rats, and by improved healthcare until the disease ran its multi-year course.

The Spanish influenza pandemic killed 50 million people worldwide during 1918-1919, more than 2.5% of the global population. The pandemic is thought to have started in the United States and spread to the rest of the world as far as the Pacific islands and the Arctic. Directed research over time that continues today has resulted in the development of vaccines to deal with the old and new strains of influenza. To the present, the vaccines have been effective in preventing a deadly pandemic as happened in 1918-1919. However, even with vaccines available, 250,000 to 500,000 people worldwide die annually from influenza.

In today's societies worldwide, the HIV/AIDS epidemic, first defined in 1981, has killed at least 25 million people (0.4% of the global population). The greatest number by far has been in sub-Saharan Africa. In 2007, 2 million people died of the disease globally (1.5 million in sub-Saharan Africa), 33 million were living with the disease, and 2.7 million people contracted the disease. Although a grand effort has been made to educate people about safe sex with respect to HIV/AIDS, the number of people contracting the disease annually continues to increase. Human biotechnology has provided therapies via medication cocktails that have significantly extended the lives of those afflicted with the disease that have access to the drugs. In recent years, low cost generic equivalents of brand drugs has given access to many in less developed countries who otherwise could not afford them. Yet, globally, only about 30% of the HIV/AIDS afflicted citizens are now receiving the anti-retroviral medicines. Research on vaccines that can reduce the possibility of HIV/AIDS infection spreading is ongoing.

Averted Threats to Major Population Losses by Disease

Recently, the world had a scare from the highly patho-genic avian influenza A H5N1 virus which was first identified in 1997 as the cause of an animal (poultry) outbreak in Asia, Africa, the Pacific, Europe and the Near East. The spread of the virus to humans was by direct contact with sick or dead poultry or wild birds or from visiting live poultry markets. Millions of birds in poultry flocks were killed and incinerated or buried at farms where public health officials identified cases of avian influenza. Other healthy flocks were vaccinated against the dis-ease. To mid-2008, there have been 387 confirmed human cases of avian influenza with 245 deaths, mainly in Indonesia and Viet Nam for a mortality rate higher than 60%. The person-to-person spread of this virus has been rare. However, the fear of a pandemic such as the Spanish influenza is a global concern. Research and development to discover a vaccine to protect hu-mans from the A H5N1 virus is ongoing.

More recently, Asia provided the world with another health crisis that potentially could have developed into a pandemic: SARS (severe acute respiratory syndrome). The cause of the dis-ease is a coronavirus (SARS-CoV). The virus spreads by close person to person contact through respiratory droplets ejected (up to 3 feet) during coughing or sneezing, or by contact with surfaces on which the SARS-CoV has settled. The first outbreak of SARS was likely in Guangdong, China in November 2002. However, the Peoples Republic of China did not report this to the World Health Organization until February 2003. By this time SARS had spread to 27 countries in North America, South America, Europe, and Asia before it was contained. One case of SARS occurred in South Africa and that person died. If SARS had reached sub-Saharan Africa, the likelihood of millions of deaths could have been the result because of the dearth of medical personnel and well-supplied clinics to identify, treat, and contain the disease. Fortunately for Africa and the world, this did not happen. The spread of SARS was likely facilitated by travel between North America, Europe, and Asia for business and tourism. The Chinese government did not report the extent of the disease until April 2003. The lack of openness by the Peoples Republic of China and the under reporting of the number of people infected allowed the disease to spread without world input directed to finding ways to cope with it. China apologized to the world thorough the World Health Organization for this unconscionable error in judgment. In total, 8098 people were affected globally and 774 died with the majority in both categories in China (7083 infected, 648 died). More than 50% of those who died were in the 65 or older age group. Research on vaccines to protect populations against SARS and evaluation of the effects of treatment with known anti-viral medications is continuing against a possible future outbreak of the disease.

Threat in Progress 2009/2010: The Swine Flu

A new influenza virus, A H1N1, was diagnosed in Mexico during March 2009 and spread into the United States where it was first diagnosed during April 2009. By July 2009, there were between 95,512 (Center for Disease Control - CDC) and 125,993 (European Center for Disease Prevention and

Control - ECDC) laboratory confirmed cases. The CDC reported 429 deaths versus the 667 death reported by the ECDC.
The greatest number of confirmed cases has been in the United
States with 37,246 and 211 deaths. The fatality rate of >4%
was highest in Argentina with 155 deaths in 3056 laboratory
confirmed cases through July 10, 2009. The number of people infected worldwide is probably much higher because most
people who have become ill have recovered without seeking
medical intervention. The CDC estimates that more than one
million people in the United States have or have had swine
flu. The World Health Organization estimates that more than
one million are or have been infected worldwide. On July 11,
2009, the WHO declared a global pandemic and labeled the
new strain of virus unstoppable.

Most at risk from influenza-related complications are children under 5 years old, adults 65 and older, pregnant women,
and people with certain chronic medical conditions (e.g., asthma, diabetes, obesity, heart disease, weakened immune systems).
The A H1N1 virus is highly contagious, passing from human
to human. Those afflicted with the disease are contagious one
day before they develop symptoms to 7 days after they get sick.
There are two anti-viral drugs for those who become very ill,
Tamiflu and Relenza. The vaccine is in production. Although
the rate of production was slower than anticipated, the vaccine
was available at the start of the Northern Hemisphere 2009
winter. Whether or not the A H1N1 virus becomes a major
global killer depends on national healthcare systems, the availability of and access to anti-viral medication, and the ability of
biotech facilities to produce enough vaccine for global populations. The U.S. Center for Disease Control suggests that young
children have two doses. Reaction to the swine flu pandemic
has already begun to hurt some economies where tourism has

fallen off (e.g., Mexico). This compounds the economic stress presently (2009/2010) experienced by most nations because of the global recession and the slowing of output and export of manufactured products.

ECONOMIC CRASH

The great depression did not directly affect global population numbers but is considered by some as indirectly responsible for the population loss during WWII. Because of this it deserves mention here so that we learn a lesson from history of actions that should not be repeated. The depression had its maximum impact from 1929-1932 but lasted until 1938-1939 when industries geared up as the specter of WWII began to materialize. The economic crash caused social disruption that left millions jobless, homeless, and penniless, and other with lower income. There were higher prices for necessities of life, governments received less tax revenue, and there were lower profits for business. Starvation was rare but many in America were undernourished and in a desperate survival mode. The United States foreign trade declined, especially from the collapse of agricultural exports such as wheat, tobacco, cotton, and lumber as crop prices dropped 40%-60% and production decreased. Heavy industry was disproportionately affected. Although the depression centered on Wall Street and the United States economy as overall employment reached 27.2%, the economic downturn was worldwide. The depression caused unsustainable high unemployment and hence failing economies in most developed industrialized countries (Table 1-5).

The great depression did not cause marked changes in global population. In the United States, for example, families became smaller. Between 1930 and 1940, there were 550,000 children born instead of the predicted 600,000. However, the

60 to 72 million deaths during WWII are thought by many to be linked somewhat to the depression because of the effects of the depression in Germany.

Table 1-5. Percent industrial workers unemployed during 1933 because of the great depression (adapted from a figure in www.english.uiuc.edu/maps/depression.htm).

USA	37.6%	Germany	36.2%
UK	19.9%	France	14.1%
Australia	24.2%	Belgium	16.9%
Sweden	23.2%	Netherlands	26.9%
Norway	33.4%	Canada	26.6%
Denmark	28.8		

As a result of the global economic downturn, loans from the United States to help rebuild Germany post WWI ceased and this brought the German economy to a standstill. In some nations, the depression caused political upheaval with despairing, frustrated, and desperate citizens turning towards nationalistic hate speeches of demagogues such as Hitler. This led to a rise of militaristic driven governments. Japan used the depression as an excuse to invade China and Mongolia for the natural resources they had although Japan suffered little from a diminished international trade during the depression. Hitler ceased paying reparations for WWI. He fostered a massive job creation program focused on industrial capability and the build up of the armaments industry. Hitler encouraged the social hate environment against religious, ethnic, and racial groups through the Nazi Party and fostered conditions that led to the abrogation of property rights and human rights. This ultimately led to the internment of children, women, and men from religious, ethnic, and racial groups, and their imprisonment in slave labor camps. It culminated in their murders through starvation and disease, shooting, poisoning in large gas chambers and the cremation of bodies in specially built ovens, or mass burial of their bodies in death camps. The great loss of life during WWII directly affected the world population for a decade.

POLITICAL CHANGES ADD TO STRESSES FROM POPULATION GROWTH

There are different social orders worldwide. These include rule of law, rule of religion, rule of fear, and rule of the weapon. In general, there are relatively few major forms of government. Each of them can have many subcategories or shadings that give them a unique character that may or may not benefit their citizens. When a change in government is peaceful, there is a relaxed feeling among citizens and little stress on society. This is not the case when a government is taken over by a coup, violent or not, or as a result of rebellious, violent forces. Tensions in a populace add to those existent from sustenance, social, and economic problems that are heightened by population growth. There is no promise that a new government can provide services and solutions to the pressing needs of citizens now suffering because of expanding populations. For these reasons, it is necessary to discuss and understand the main forms of government, examine downfalls of governments, and evaluate situations where there have been violent changes.

Main Forms of Government

Democracy is a multi-party form of government in which citizens' rule through elected representatives who listen to their constituents to determine what the people want. Representatives then debate the pros and cons of legislation and enact laws to fulfill citizen wants. Republicanism is in this category.

Communism is a one party system. The head of the party is the leader. The party sets the laws and controls all aspects of a society (e.g., education, healthcare, business, industry, agriculture, natural resources). In theory the system seeks equality for all citizens with one social stratum, the working class. In reality, this has not been the case. Some forms of socialism are modifications of communism.

A monarchy is a government in which a king or queen (in a few cases an emir or sultan) rules as the absolute monarch or as a constitutional monarch with a prime minister and elected parliament (national assembly, congress) setting the law. Heredity determines the ruler in most monarchies but some are elective. Others are taken by force and may be considered dictatorships.

In a theocracy, the laws are guided by religious belief based on a society's holy book(s) such as the Islamic Koran or the Christian Bible. Only the official religion is permitted. A religious leader or religious council directs the government. Some republics are theocratic but allow the practice of other religions.

Only a few people manage an oligarchic government. A theocracy as described above may be governed under an oligarchic system.

In addition, there is anarchy or full rule by no one. Similarly, some governments are ethnocentric and are intolerant of other ethnic groups living in their domains. Others are racist in nature and rule by the whip, jailing, and the sword. These include despotism, feudalism, police state, military dictatorship, fascism, and totalitarianism. They are evil systems of government that have no place in civilized societies but they do exist and ultimately will fall in the future as they have fallen in the past.

Downfall of Governments

Governments can fall when they do not meet the needs of their citizens. They may fall if their leaders and close cohorts live well and in some cases enrich themselves at the expense of the citizenry. Government leaders may ignore basic human rights and property rights in favor of family or friends. Their countries may suffer economic demise because of poor governmental planning or poorly prioritized investment, often times in the military, and also because of corruption. Changes in governments may come about by free elections, or governments may be brought down by a coup (popular or military), or by controlled or uncontrolled populist driven revolutions, and in some cases by traditional or internationally pressured negotiations. We have experienced examples of all of these situations during the past generation or two.

In a few cases, repressive governments led today by military dictatorships or by despots supported by the military, such as Myanmar (Burma), North Korea, or Zimbabwe, have not changed. They have yielded little to international pressures for change to help provide necessities for their populations. Eastern European communism (USSR + puppet states) collapsed during the period 1989 when the Berlin Wall came down and brought the "iron curtain" down with it, to 1991 when the thawing of the so-called "cold war" ended. This mind-boggling rapid change was brought on in grand part by the economic demise and socio-political pressure from citizens of the communist-controlled societies. The Federation of Independent States that was the former Soviet Union has taken on the mantle of republicanism but in different systems, some of which retained the trappings of communism and dictatorship and other which became democracies. China and Cuba persist with their form of communism in which free market systems are allowed to operate but are largely controlled by the state.

Examples of Violent Change

During the 1970s, Idi Amin Dada led a coup and became the military dictator of Uganda. He fostered human rights abuses, political repression, and ethnic persecution. He approved militia killings and expulsion of Asians (mainly Indians) from the country. Popular dissent and an invasion of Uganda by Ugandan rebels and the Tanzanian army led to his downfall and exile to Saudi Arabia where he died in 1993. Uganda is now governed by a president, prime minister, and national assembly but still has political turmoil because of questions as to how state funds are used.

Similarly, during 1973 Augusto Pinochet, the Chilean army commander-in-chief, engineered a coup of a freely elected (socialist) government. A military junta suspended the congress in 1974, functioned as a legislative body, and made Pinochet president. He acted as a dictator and eliminated political party participation in the government. He repressed public dissent and condoned the killing of thousands of dissenters. Subsequently, under international pressure and increasing pressure from the Chilean citizenry, political activity was again allowed. A referendum to hold free elections was held and won by the opposition that deposed Pinochet in 1988. In 1989 there were free presidential and legislative elections and Chile was once again a democracy. Pinochet was later tried for crimes against the state and died while under home detention.

Also, during the 1970s, a military junta took over the governance of Argentina. There was suppression of political activity and the state/military-directed killing of tens of thousands of dissenters and political activists. In 1982, with the probable underlying purpose of diverting public attention from a failing economy, the junta initiated a war with Great Britain to take over the Falkland Islands (Islas Malvinas). The military lost the war decisively, was largely discredited, and acceded to elections for a democratic civilian government. In 1983, free elections were held to elect a president and congress. The country has been a democracy since then although it has gone through some terribly difficult economic times. Military leaders have been taken to trial, convicted, and sentenced for crimes against the state. Changes in Argentine governments will continue to take place in a democratic environment with the avowed purpose of legislating and acting for the benefit of the citizenry.

South Africa presents a special case in which violence during the reign of an apartheid government brought about a move towards negotiation to end apartheid. The negotiation forced upon the apartheid regime was the result of internal pressures (the violence and human rights activists) and international economic, political, and social pressures. This brought about free elections. Those who suffered from the deposed regime formed a government that has been, to a good degree, fair for all the citizens.

AGE-GENDER POPULATION GRAPHICS
The distribution by age and gender in a population is represented by what is traditionally called a population pyramid. To prepare an age-gender graphic, the number of males in a population is plotted against the number of females in a nation in age categories starting with 0-4 years of age and then

in 5-year increments from 5-9 years old to 95-99 years old. In recent years, demographers had to add a 100+ increment for some countries. When age-gender distributions were initially graphed, populations in most countries showed larger numbers of young persons for both males and females and gradually lesser numbers in categories that represented older people. For the purpose of planning, the population pyramid can be divided into four grouped segments. These are the numbers coming in to childbearing age (0-15), those in the childbearing group (15-49), in the older adult working group (50-65), and then the retirees and the aged (65-100+). The shape of the distributions that were initially graphed (growing populations) was that of one side of a pyramid, hence the designation as population-pyramid. Nations use the data in population pyramids to assist planners to prepare for the needs of future populations whether they are growing or contracting. These include sustenance (food, water), housing, schools, infrastructure, energy, healthcare, employment, care for the aged, and other parameters necessary to sustain a society with a good quality of life.

Population pyramids for all countries and major world regions are derived by the U.S. Census Bureau from an International Data Base.[9] They can be called up for all years through 2050 at www.prb.org at Data Finder mode. These graphic representations can change as demographers receive yearly data from each country.

Examples of Population Pyramids

Continued Growth
The population pyramids for Ethiopia for 2010, 2030 and 2050 (Figure 1-1) are illustrative of a population that is expanding with increasing numbers of persons coming into

childbearing age as well as large populations in childbearing age. The Congo (Kinshasa, Figure 1-2) and Nigeria (Figure 1-3) have population distributions similar to Ethiopia but the Congo (Kinshasa) shows promise of slow down of growth for the 0-15 age group whereas Nigeria shows a marked decrease in numbers of the 0-15 age group. The population pyramids for India (Figure 1-4) illustrate a populous country that, similar to Nigeria, is on the path towards a significant reduction in the rate of population increase. The Indian fertility rate of 2.7 in 2008 is projected to drop to 2.2 by 2025. This not withstanding, India will overtake China as the most populous country by 2035.

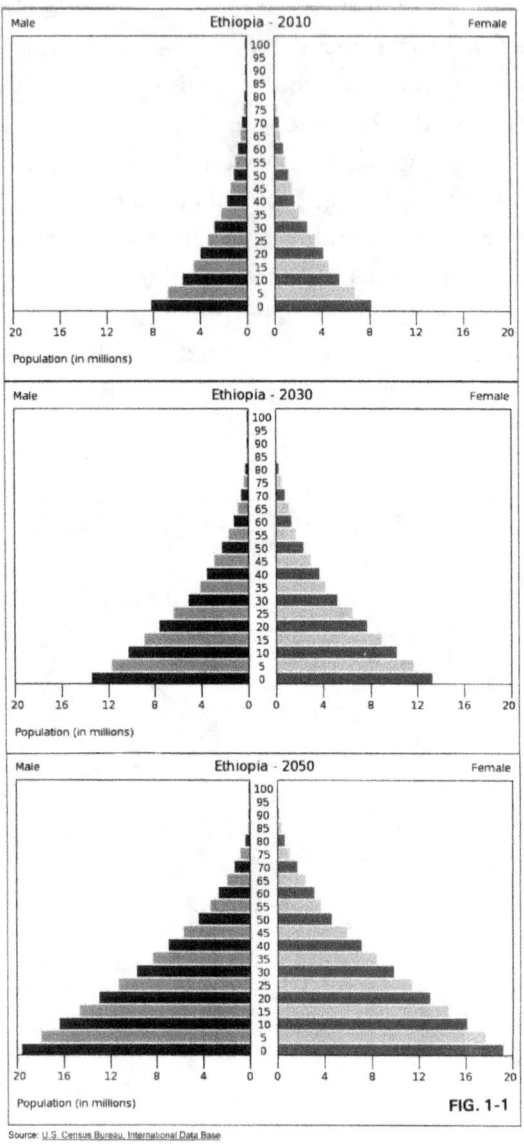

Figure 1-1. Population pyramid summary for Ethiopia. U.S. Census Bureau, International Data Base, 2009.

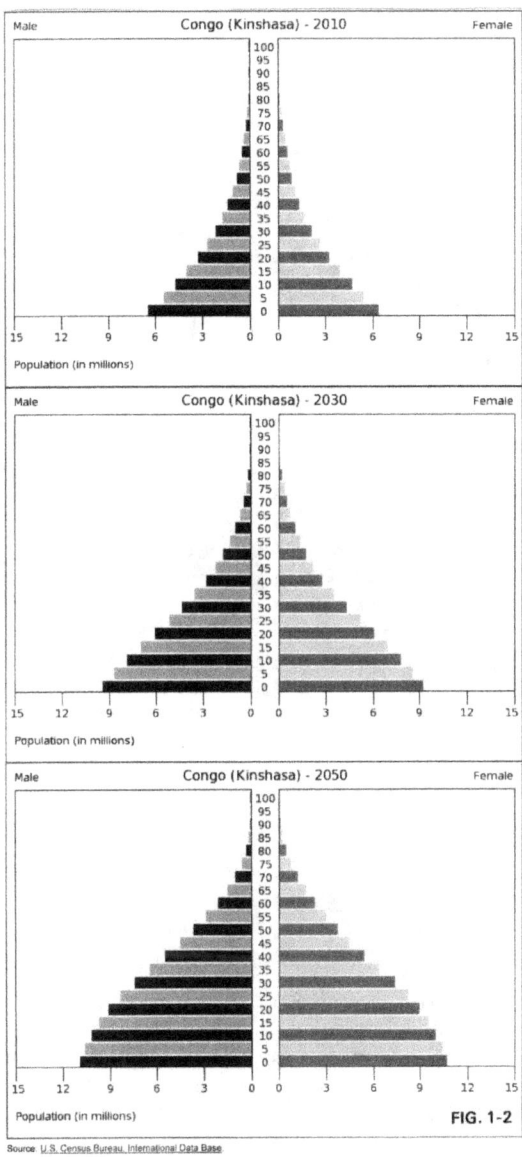

Source: U.S. Census Bureau, International Data Base

Figure 1-2. Population pyramid summary for Congo (Kinshasa). U.S. Census Bureau, International Data Base, 2009.

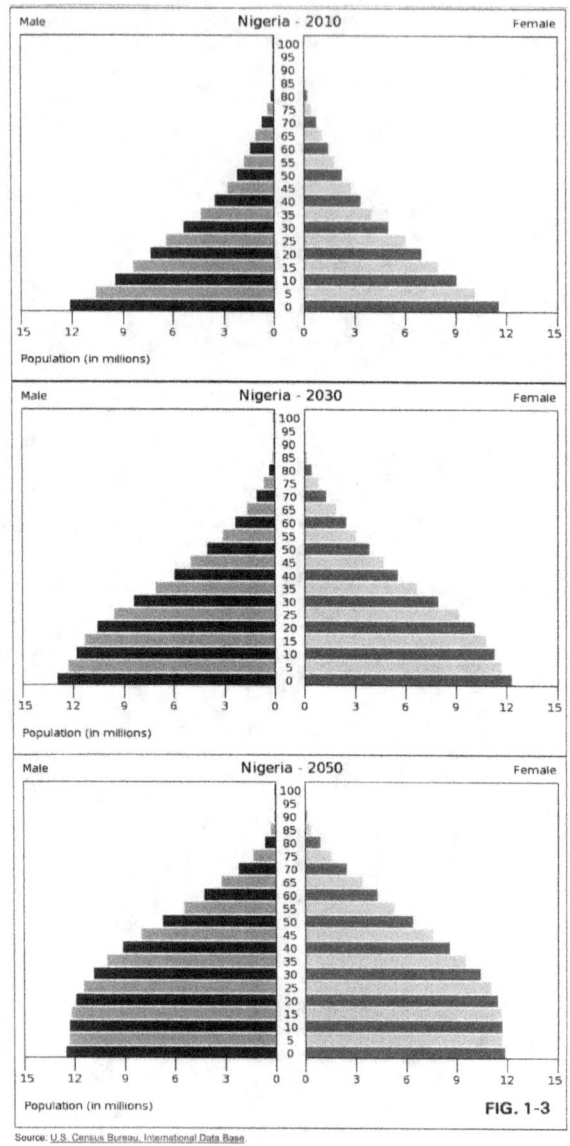

Figure 1-3. Population pyramid summary for Nigeria. U.S. Census Bureau, International Data Base, 2009.

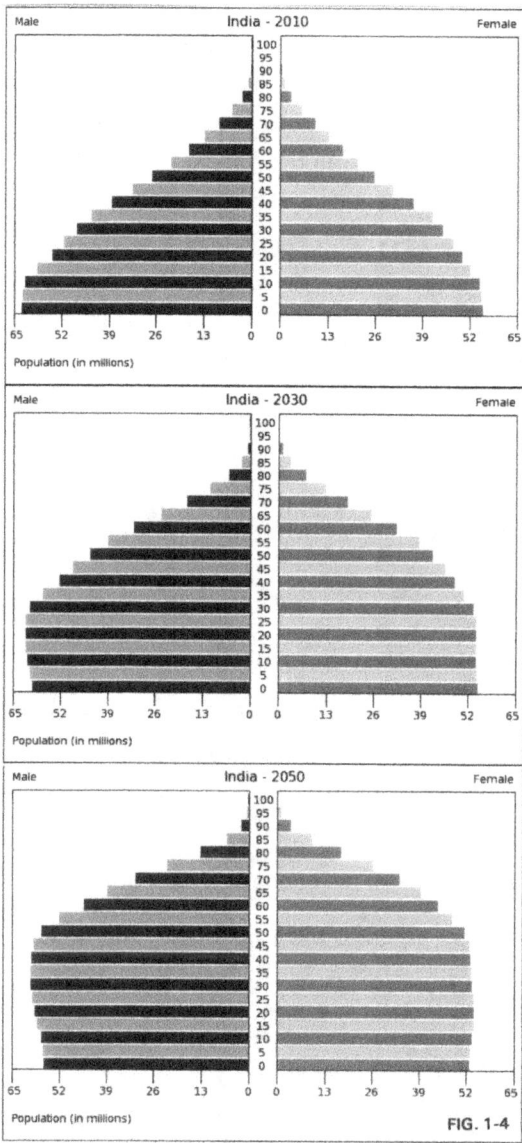

Figure 1-4. Population pyramid summary for India. U.S. Census Bureau, International Data Base, 2009.

Towards Stabilization

As populations grew and aged, and as fertility rates of some nations lessened to slightly below replacement levels, the shape of the graphic changed. A rather stable population had a more cylindrical form somewhat pinched in the 0-15 and 15-49 age categories. This is illustrated in the population pyramid sequence for the United Kingdom (Figure 1-5).

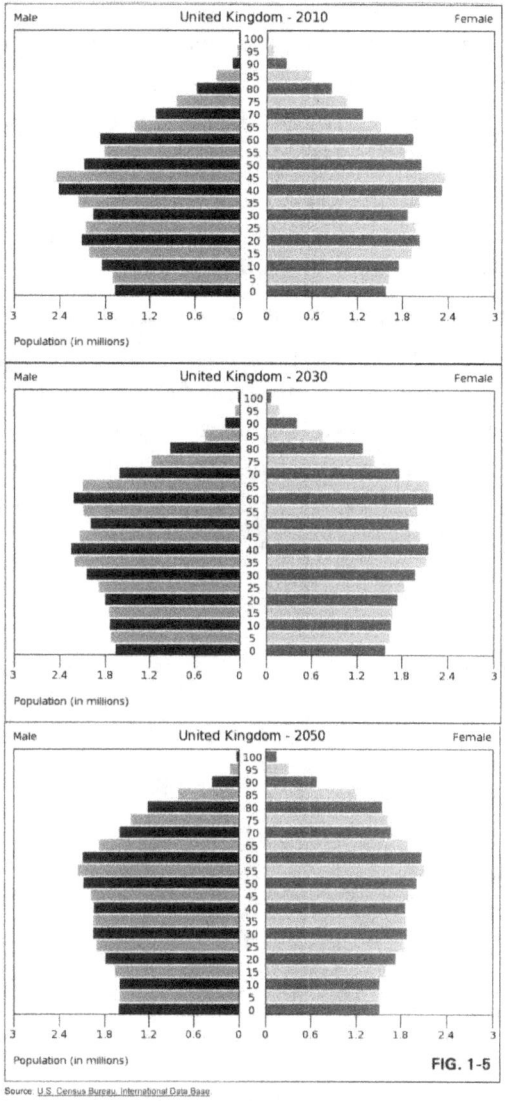

Figure 1-5. Population pyramid summary for the United Kingdom. U.S. Census Bureau, Internation Data Base, 2009.

Contracting and Aging Populations

Finally, as fertility rates in some nations dropped well below the replacement level and populations contracted, the shape of the graphic assumed that of a pedestal with a markedly pinched base. There was a notable decrease in population in the 0-49 age categories, a bulge in the 50-80 age categories, and a tapering off in the numbers of older citizens. Japan shows this type of age-gender graphic and represents an aging and contracting population (Figure 1-6). The population pyramid summaries for Russia (Figure 1-7) and Ukraine (Figure 1-8) are similar to that of Japan but without the gradual shifts in distributions that Japan shows. China has a population distribution (Figure 1-9) similar to those of Russia and Ukraine, certainly influenced by the one child per family policy (for urban populations) that has been in place for several years.

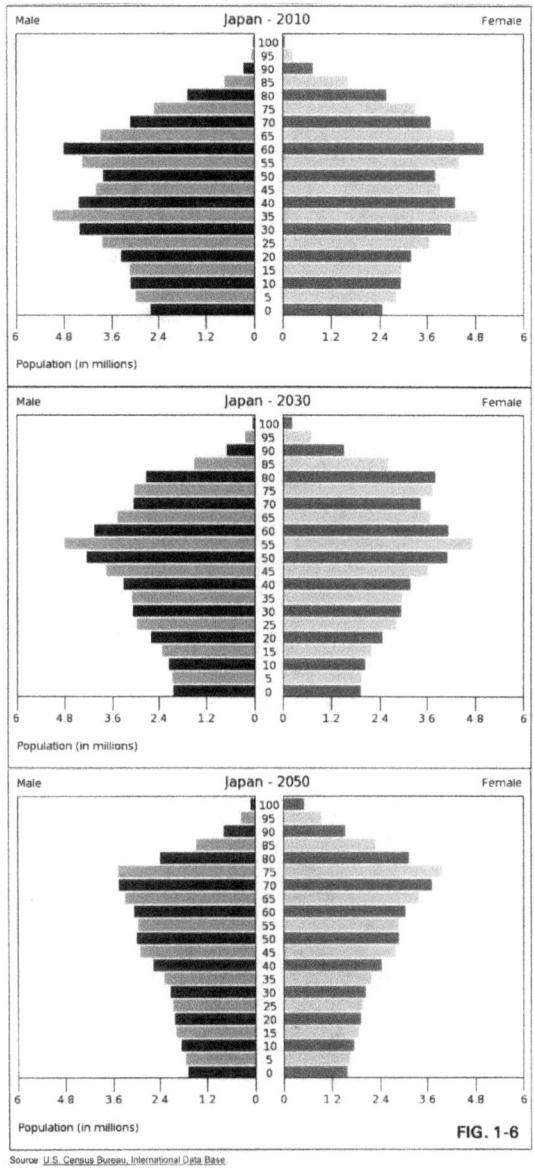

Figure 1-6. Population pyramid summary for Japan. U.S. Census Bureau, International Data Base, 2009.

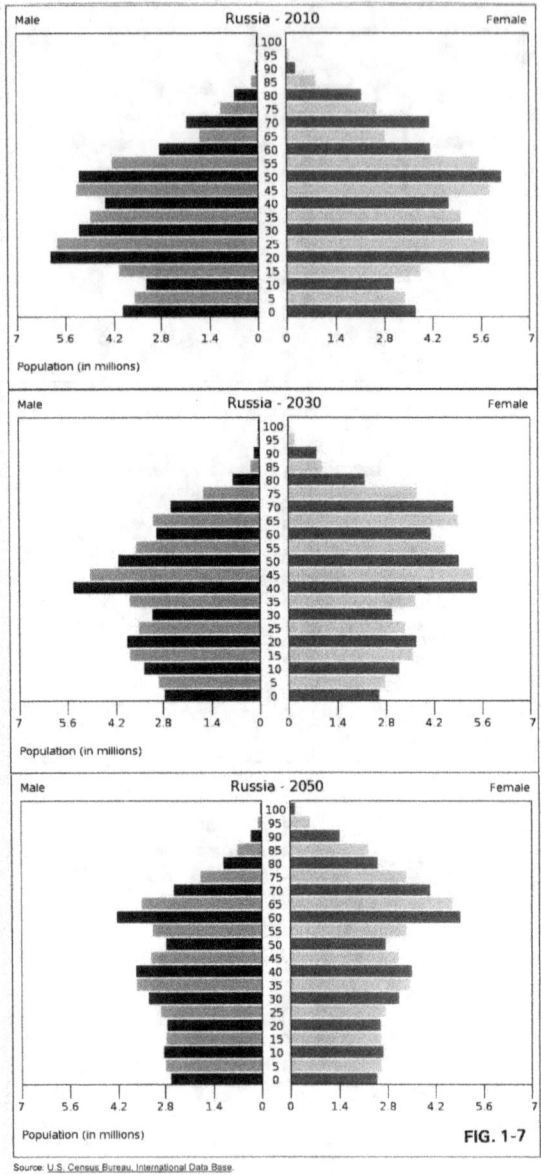

Figure 1-7. Population pyramid summary for Russia. U.S. Census Bureau, International Data Base, 2009.

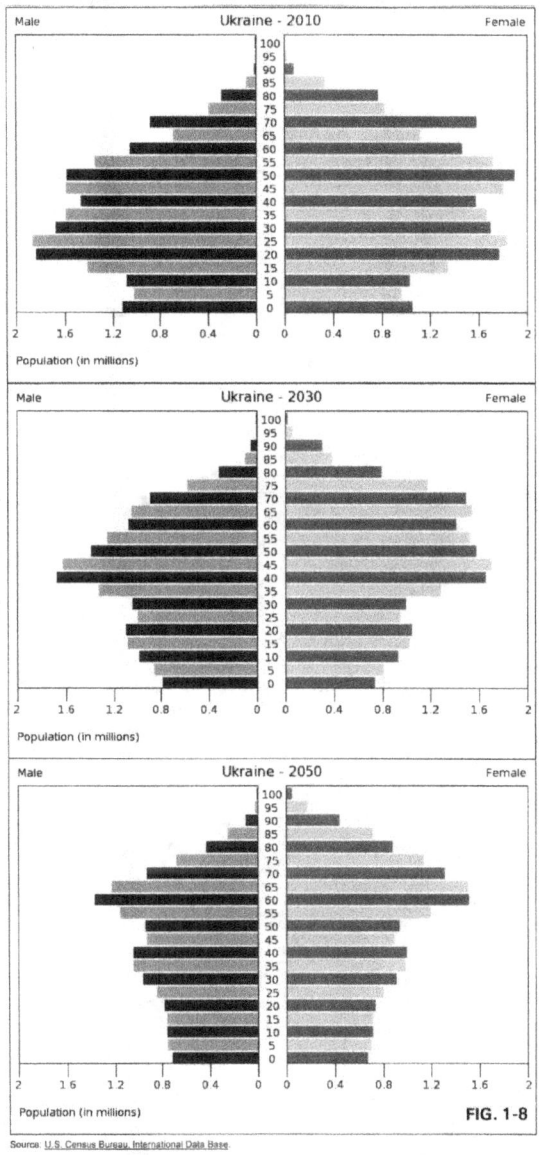

Source: U.S. Census Bureau, International Data Base

Figure 1-8. Population pyramid summary for Ukraine. U.S. Census Bureau, International Data Base, 2009.

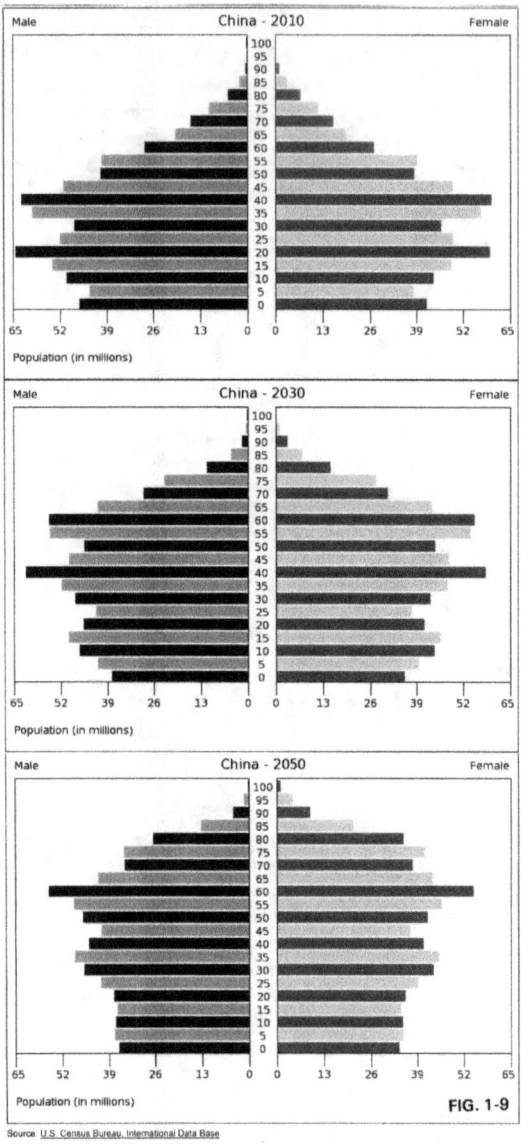

Figure 1-9. Population pyramid summary for China. U.S. Census Bureau, International Data Base, 2009.

Trends for the United States, China, India, Africa

The population pyramids for the United States show a trend towards stabilization for the year 2010 but has trends towards growth for 2030 that increase for 2050 (Figure 1-10). A part of this growth is due to legal and illegal immigration and the tendency of immigrant populations in the country to have higher fertility rates. The fertility rate in the United States for 2008 is 2.1 and it could rise to 2.2 by 2025. In some estimates, up to 40% of population growth may be from legal and illegal immigrants and their offspring. A similar situation may affect populations in Europe such as in France and the Netherlands if they do not manage the number of immigrants seeking entry to a country. Most immigrants come for economic opportunities but also because their human rights are protected, because there is ready access to social services (e.g., healthcare, education), and because governments tolerate political activity that in some cases are anti-governmental.

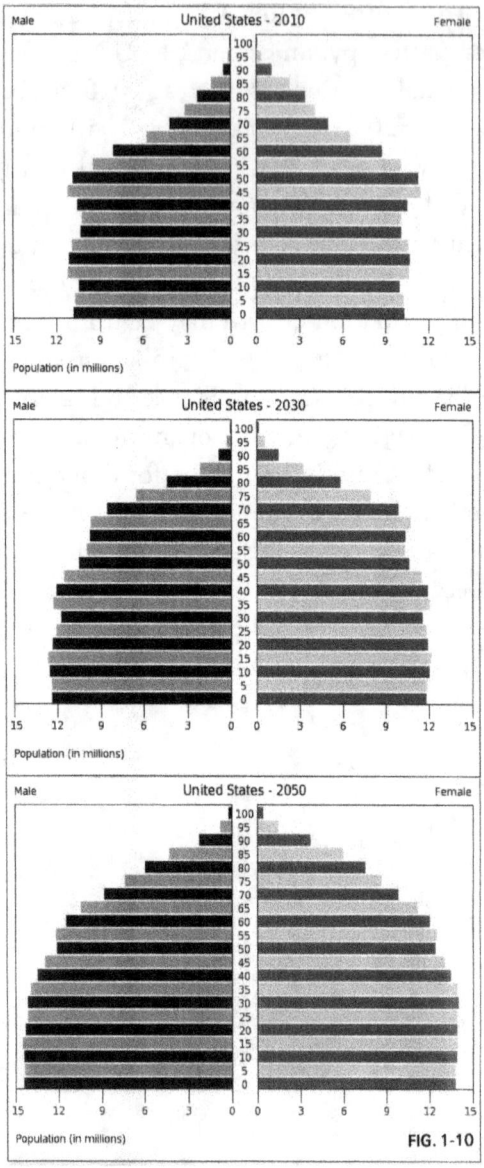

Figure 1-10. Population pyramid summary for the United States. U.S. Census Bureau, International Data Base, 2009.

As we assess global conditions with respect to today's population, through to 2030 and on to 2050 as illustrated in Figure 1-11, it is clear that China, India, and especially the African continent will contribute most to global population growth.[10] A majority of less developed and developing countries have increasing populations but not at the absolute numerical levels of China, India, and Africa. Population growth in the more developed countries tapers off during 2020 and remains stable through 2050.

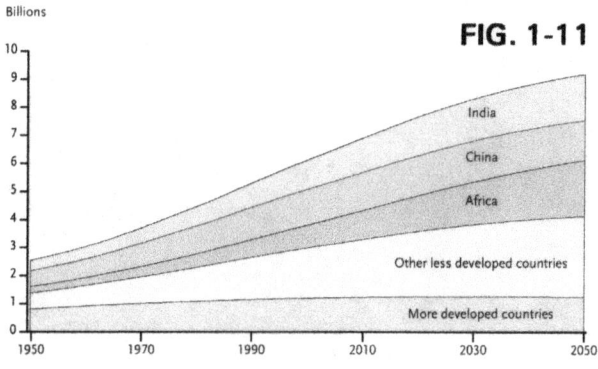

Source: UN Population Division, *World Population Prospects: The 2006 Revision*, Medium Variant (2007).

Figure 1-11. Developing regions and developing nations take up an increasing share of the global population (after the United Nations Populations Division, 2007).

The relation between what may be expected in population distribution in the more developed and less developed/developing countries is strikingly evident in their population pyramids. Figures 1-11 and 1-12 show respectively that there is an expanding young population in the less developed countries and that for 2010, more developed countries have fewer young people and less developed countries have more young people.[11] This latter situation does not bode well for the fu-

ture of global citizenry that depends on the stabilization of
those segments of the populations coming into childbearing
age. If stabilization is not achieved, the world population may
grow to a number of inhabitants that is not supportable by
ecosystem Earth. Some demographers determine this number
to be 9.5 billion people. Others propose a figure of 7-8 billion
inhabitants.

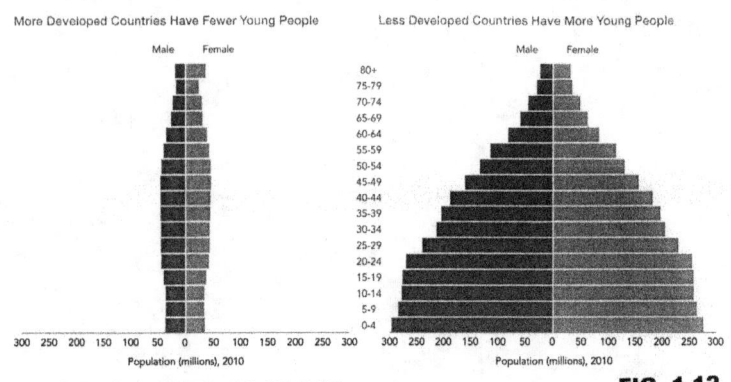

Figure 1-12 A comparison of numbers of elderly people relative to young people in more developed countries and in less
developed countries (after the United Nations Population
Division, 2007).

AFTERWORD

Books, academic papers, and solicited reports from experts
and think tanks commissioned by international organizations,
governments, and the public, underscore the threat to humankind from expanding populations (e.g., sources of nourishment,
natural resources). The results from these have been concerted
global efforts to reduce fertility rates where necessary and hence
the population growth rate.

Education, family planning, and the availability and use of contraceptive methods, and incentives by governments globally for citizens to use them have effectively and rapidly reduced fertility rates and rates of population increase in many countries. This is what is needed if humans are to avoid the suffering such as from the pangs of famine and starvation, from debilitating and deadly diseases carried by contaminated water and food, or from vector-borne maladies and yet unknown viruses. This is what is needed to reduce the threat of local, regional, and even global wars to get access to the necessities of life. This is what is needed to prevent pandemics that in their own ways could conceivably reduce populations in some countries to sustainable levels.

There are decreasing rates for global fertility and population growth. Most nations are in the slow down mode but a small number are not. The rate of decrease in fertility and population growth has to be expedited if the Earth population is to stabilize at a projected 9.5 billion inhabitants or less. At this number, advances in technology may develop the means to at least provide sustenance (water, food) for global communities. This is questionable because technology has not advanced quickly enough to be able to achieve that purpose in 2008 for billions of people worldwide.

War and disease will cause population losses. However, as described earlier in this chapter, most are temporary aberrations that are more than made up for by global births. Only a "super nova" event similar to the black (bubonic) plague could really cut back the world's population. Certainly, a global nuclear conflagration could do the same. However, a war would likely be focused on targeted countries or regions leaving the largely unaffected nations to "pick up the pieces" to nourish and heal the Earth and in their own ways be nourished and healed by it.

Changes in government policies on prioritized invest-ment and global investment in research and development of technological advances that improve supply and distribution of those commodities that provide a good quality of life are desirable. When achieved, such advances should allow us to sustain socially stable, productive and healthy populations. We have not done this for perhaps 1/3 of today's 6.8 billion Earth inhabitants. It is difficult to conceive of being able to do this in a two generations future for 9.5 billion people or more. The limitation to global population growth is the car-rying capacity of the Earth. This is the planet's capability to provide what is necessary to sustain life and nurture the hopes and aspirations of its inhabitants. If this capacity is exceeded, there will be population crashes in countries least able to cope with them. Small populations of the strongest people (e.g., healthy, self-reliant, economically-advantaged) will survive but the rest will die. Stresses on governments, ecosystems, and current populations in dealing with the realities of what to do now for deprived humankind and what can be done to provide for future expanded populations are explored in the following chapter.

CHAPTER 2. STRESSES EXERTED BY EXPANDING POPULATIONS ON GOVERNMENTS, ECOSYSTEMS, AND EXISTING POPULATIONS

STRESS

Stress as used in this book is a biological term. Stress is the psychological result of internal (physiological) reactions to external stimuli (forces). Stress develops from the failure (or inability) of humans to adapt to change or to respond expediently to emotional or physical threats whether actual or imagined. A response may be to fight for what one deems right or to flee from the real or perceived threat. Animals show the same reaction. Stress (a.k.a. pressure or tension) is triggered by one or more physical, biological, economic, or socio-political changes that negatively impact environmental vitality and stability.

Stress from population growth compels governments to plan ahead to do what is necessary to supply the basic needs for a greater number of people. If this is not done, there will be tensions among the populace and between nations. This will threaten the viability of ecosystems because of factors such as encroachment, overuse, and pollution.

Stress can be brought about simply from trying to meet the challenges of everyday life and adapt to them. It can originate from concerns about food, housing, health, education, freedom, mobility, or unemployment. Noise, crowding, traffic congestion, extremes of temperature, over-illumination, or threat of terrorism also cause mental strain in urban/suburban settings.

Stress can involve changes in environmental conditions during the constant physical and behavioral interchanges of humans with ecosystems. These might be pollution, natural or anthropogenic hazards that are recurrent such as drought or shocking such as an earthquake or a volcanic eruption. There may be shrinking ecosystems from human encroachment or loss of protective environments such as from excavation of wetlands or leveling of coastal sand dunes. Clearly, many factors, some related, some not, can rouse states of stress at low, moderate, or intense levels.

The stress process starts with a state of alarm and adrenaline production, a short-term resistance as a coping mechanism, and ends with mental exhaustion. In humans, stress reveals itself by a broad spectrum of emotional, behavioral (psychological) and biological/physical (physiological) symptoms and may manifest itself biologically and/or physically. Common symptoms of stress in humans are irritability, muscular tension, inability to concentrate, and a variety of physical reactions (headache, accelerated heart rate). There is a link between stress and disease (e.g., ulcers, depression, trouble with the digestive system). This book emphasizes the stressors on societies (existing populations), governments, and ecosystems that originate from growing populations (or in some cases, contracting ones).

Stressors from Growing Populations

Stressors from expanding populations on the very populations that fueled the growth may be felt directly or may have indirect effects. These are presented in Table 2-1.

Table 2-1. Direct and indirect stressors caused by expanding populations. These stressors put pressures on or cause tensions in governments, ecosystems, and contemporary populations.

Biological and physical - from less water and/or polluted water (chemically, biologically)

- from less food, poor quality (less nutritious) food, tainted food
- from bad air locally, regionally, globally
- from lack of sanitation controls, if any
- from encroachment on forest land and agricultural terrain causing
- a loss of habitats in host ecosystems

Economic - competition for natural resources based on supply and demand

- limited and focused investment, if any
- little employment possibilities or underemployment of capable cadre
- minimal acquisitive power beyond essentials, if that
- no training or retraining programs
- poor infrastructure development or decaying infrastructure

  no loans for funding to improve or create infrastructure
- limited communication networks
- small business loans may not be existent

Sociological - increased population density and less green space

- insufficient or poor quality housing
- limited availability of healthcare in general and local clinics in particular
- limited access to cooking/heating fuels
- limited availability of electricity
- poor availability of childrens' educational facilities, teachers, and tools

Political - free vote, fixed vote, no vote
- no elected government
- family, clan, tribal, regional, national loyalties
- lack of personal security
- no habeas corpus
- no free speech
- no right to assemble
- no respect for human rights
- loss of property rights
- non-independent judicial system

ECOSYSTEMS

Earth is the master ecosystem and is comprised of a vast array of macro- and micro-ecosystems in the atmosphere, in water (terrestrial, estuarine, marine), and on the land. Each ecosystem has many parts that influence the continual flow of energy and cycling of nutrients that sustain its vitality and productivity. These components fall into four general categories: biologic, geologic, hydrologic. or atmospheric. The biologic element includes trees and other vegetation, animals (+ fish and foul), insects, and microorganisms. Rocks, soils that form from them, minerals (+ ores), and many and varied landforms (e.g., mountains, hills, plains, canyons, valleys) comprise a geologic component. The hydrologic factor includes surface and subsurface (aquifer) water, rain, snow, ice, and moisture (in air and in soils). Sun, wind, and other factors that influence weather and climate (e.g., gases, aerosols, particulates) comprise the atmospheric element. Any change or interruption in one or a combination of these influencing factors that reduces nutrient cycling or disrupts food chain/food web continua in ecosystems will stress them. Change or interruption will diminish ecosystems' viabilities and abilities to sustain the well-being of today's populations and future generations of humans and other life forms ecosystems nurture.

Ecosystem stressors may be local, national, regional, and global. Thy can be natural or influenced directly by humans, or indirectly as the result of anthropogenic activities. For example, population growth as a global stressor is a natural process. Man-made global ecosystem stressors include pollution from mercury (Hg) in the atmosphere, thinning of the ozone layer, and climate change from global warming. Extreme natural events such as massive volcanic eruptions, strong earthquakes, category 3-5 hurricanes (sustained winds of 111-155 mph), and widespread flooding can be ecosystem stressors at local and regional levels. Air and water pollution stresses ecosystems in all geographic regimes. For example, air pollution may be from acid rain, Hg, other potentially toxic heavy metals and toxins, and particulates at locations in proximity to and downwind of smelters, coal-fired power plants or other toxin emitting industrial sources. Similarly, locations down flow or in proximity to water sources can become contaminated and present risks to ecosystems and their inhabitants. The water pollution can be nutrient runoff from feedlots, nutrient and pesticide runoff from agricultural terrain, acid mine drainage from abandoned or working operations, and polluted runoff from dumps or poorly engineered waste disposal sites. In addition, vector-borne diseases (e.g., transmitted by insects or rodents) can originate from waste disposal sites.

Certainly, inadequate planning for land use and resources extraction or resource use can stress local ecosystems and become a national problem when pollution plagues many such systems. In some terrestrial and aqueous environments, the planned introduction of species that become invasive, or invasive species from unidentified sources can stress the systems by interfering with natural food chain/food web pathways. There is no end of ways that ecosystems can be stressed so as to lose major sources of their vitality. This is detrimental to life forms that depend upon them for sustained yields of natural resources or uses. The stressors can be influenced by anthropogenic activities and have profound socio-cultural and economic consequences for humans and their welfare.

Humans stress ecosystems to the limit. The World Resources Institute reported that 1/2 of the world's wetlands were lost during the 20th century, that logging and land conversion have reduced forest cover by at least 20% and perhaps by as much as 50%, and that about 9% of the world's tree species are at risk of extinction. The Institute further stated that nearly 70% of the world's major marine fish stock were over harvested or are being fished at their biological limit, and that in the last half of the 20th century soil degradation (e.g., from erosion, nutrient depletion, salination) has affected 2/3 of the world's agricultural land.[12] The continuing degradation of Earth environments does not bode well for today's global populations and their growing numbers that may stabilize at 9.5 billion people but may just as readily increase to higher numbers.

EFFECTS OF STRESSORS: ON THE POPULACE, GOVERNMENTS, AND ECOSYSTEMS

Stressors as presented in this book are arbitrarily categorized as physical, biological, economic, and socio-political. They are inexorably linked because stressors in either category beget others in all categories.

Biological Stressors

Biological stressors (lack or water and/or food) are essential commodities without which a society cannot thrive, or at the worse, survive. For example, humans can go without water for about 4 days or without food for about 40 days during which organs start shutting down and death ensues.

Water - Of the approximately 6.8 billion people on earth in 2009, about 1.2 billion live in so-called developed nations and about 5.6 billion live in less developed and developing nations. Safe water is provided to most of the populations in developed nations. About 1.2 billion people, or more than 21% of the populace in less developed and developing nations do not have access to safe water and 2.7 billion people (almost half), do not have water for sanitation.[13] They suffer from this lack of sufficient water through an array of sicknesses and by death. In addition, without enough water for irrigation, yields of food crops shrink, food quality diminishes, and a cycle of malnutrition can take root.

In theory, the earth has enough water for human and ecosystem needs. In reality, the water is irregularly distributed nationally and regionally. Thus, there are water-rich districts and water-poor districts. Where there is an infrastructure to move water readily from where water is plentiful to where it is in deficit, populations can flourish. Where there is no infrastructure to do so, populations are in distress. In many cases, water that may be available to people is contaminated. If it is used for drinking, cooking, personal hygiene, or irrigation, ingested or absorbed contaminants via food or water can cause grave sicknesses. If not treated quickly and properly, such sicknesses can terminate in death (vis a vis the 2008/2009 cholera outbreak in Zimbabwe - see Sanitation section later in the chapter).

Food - As noted in a previous paragraph, without sufficient water to sustain irrigation for food cropping and for animal husbandry, populations cannot feed themselves. Although there is enough food produced in the world to feed all of its population, food transport and its distribution to needy, less developed regions can encounter bottlenecks. Often times this is the result of inadequate infrastructure (e.g., impassable roads, bridges in need of repair) or political and military interference. As a consequence, there is malnutrition that can evolve into famine and starvation. In the developed world, there is no famine and only a relatively small number of people suffer from malnutrition and starvation. In the developing world, more than 3.5 billion people suffer from mild to acute stages of malnutrition where there is little food, poor quality food, or famine and not enough food transported from where it is in surplus to where it is needed. Of this number, almost 1 billion suffer from chronic starvation.[14] This is in real time 2009/2010, and portends badly for many countries where populations are projected to grow up to 50% in a few generations.

Air - The global atmosphere is not pristine. It may contain heavy metals, particulates, and toxic gases (e.g., SO_2 reacts to form sulfuric acid, the main component of damaging acid rain). Globally, air pollution contributes to the deaths of 3 million people annually. The amounts of atmospheric pollutants vary depending on location with respect to urban or rural areas and industrialized or agricultural zones. Respiratory and other health problems can develop from the ingestion and bioaccumulation of one or a combination of atmospheric contaminants in a body over time. In the case of Hg, for example, large food fish (e.g., swordfish, tuna, king mackerel, shark, tilefish) accumulate and bio-magnify Hg during their life span. When these fish are harvested and eaten regularly, the toxic metal bioaccumulates in consumers' bodies. This presents a long-term health risk particularly for pregnant women and small children.

Obviously, populations of humans and other life forms located near and downwind from major industrial sources of air pollutants are at high risk from pollutant ingestion by respiration. Similarly, urban populations close to industrial zones and with heavy vehicular traffic, especially in topographic low areas that can undergo atmospheric inversions, are at a health risk from breathing smog. Smog is a lethal mixture of ground level ozone, other gases, and particulates that accumulates and lays in close to the ground surface. When the smog hangs in an area for days, respiration of the toxic mixture can cause bronchial and cardiovascular problems. Thousands in London, Teheran, and other densely populated urban centers died when outdoors and being exposed to smog for a relatively short period of time (e.g., four days in London, 1952).

The density of citizens in urban/suburban centers will increase with population growth. Peoples' needs for industrial and manufactured products, transportation, and electricity, for example, will surely cause more pollutants to emit to the atmosphere unless best available pollution capture equipment is installed and used at emitting sources worldwide. This requires also that captured pollutants be recycled for safe use or securely be disposed of if they are not recycled.

Sanitation - Poor sanitation and illness in populations is prevalent where there is a lack of water to move sewage away from human habitation. The preferable move is to collection and treatment facilities, or, where there are no such facilities, to areas away from population centers. Diseases linked to sanitation failings can arise and spread quickly among a population. They stress citizens physically and psychologically. These diseases are often triggered by interruptions to a treated water supply. Overcrowding aggravates the health situation. The lack of water and soap to wash hands after using toilets is one reason for the occurrence and spread of diarrhea diseases. Unsanitary conditions contribute to more than

5 million deaths annually. More than half of these deaths are children.[13]

A major cholera epidemic broke out in Zimbabwe during August 2008. It affected both urban and rural populations. Large cholera outbreaks in the country during 1999 and 2003 were contained but they signaled the probability of a future major epidemic of the disease. The earlier outbreaks were attributed to contaminated water because of broken pipes and a lack of chemicals to treat water. These events signaled problems in the water system but were not heeded by Zimbabwean health and sanitation officials. They knew that the water treatment and sanitation infrastructure was failing. They knew that cholera is transmitted through contaminated water and food and is highly infectious.

With few doctors, closed hospitals or hospitals that were woefully understaffed and without a stock of medicines to effectively treat cholera patients, the 2008 epidemic was not contained. By December 10, 2008, the Zimbabwe Ministry of Health recorded that 15,572 persons were infected by cholera with 746 deaths for a mortality rate of <4%...as high as 20%-30% in some remote villages. On December 12, a defiant and disillusioned President Mugabe insisted that there was no cholera disease in the country. By December 27, 1518 people had died of cholera of 26,497 suspected cases of the disease (5.7% mortality). On January 23, 2009, the World Health Organization (WHO) reported that there were 50,000 reported cases of cholera and 2773 dead with the mortality rate steady at 5.7%. The progression of the disease is the product of a collapse of the healthcare system in a failing state. World Health Organization assessors believed that these figures were under-reported. They warned early that more than 60,000 people could be sickened by the cholera with as many as 6000 dead. The number of cases soared to 76,127 with 3623 dead by February 17, 2009 and then to 89,018 with 4011 dead by March 9, 2009. The pace of the epidemic seemed to be easing. By May 26, 2009 the number of cases of cholera was 98,424 with 4276 dead from the disease (<4.3% mortality rate). This was the disease containment point as evidenced by only a slight change in cases reported on July 15, 2009 of 98592 infected and 4288 dead.[15] With proper treatment of early re-hydration using IVs and oral re-hydration salts, most patients can recover from the disease and the mortality rate would likely be 1%-2%. Spread of the disease can be arrested by using proper food safety practices and good personal hygiene, both of which are limited in Zimbabwe because of the lack of safe water, a consequence of a failing state. Safe water was trucked in by the WHO and collaborating outside groups that have come to help the Zimbabwean people. Millions of water purification tablets have been distributed as a short-term solution to purify contaminated water. The Zimbabwean government reluctantly accepted help from the WHO and other government and NGO groups in an effort to contain the epidemic and treat those who carry the cholera disease. There have been confirmed cholera cases where Zimbabweans seeking treatment for the disease carried it into the border towns of Musina (South Africa), Palm Tree (Botswana), and the Guro district (Mozambique). Because of this grand social health failure, grave political problems, corruption, economic depression, and heavy-handed repression of political opposition groups, many categorize Robert Mugabe as a despot and Zimbabwe as a failed state.

Economic Stressors

It is evident that when there are too many people and too few jobs, there is economic distress. This can be exacerbated when unscrupulous employers pay less than minimum wages with no benefits when they do hire employees, especially undocumented persons. There is economic distress when there are no positions in careers people have prepared for so that they can only get menial jobs. Investors are unwilling to deal with a lack of educated cadre to support job-creating projects so that the state must underwrite training or retraining programs. To attract investment in large-scale job-creating projects, there must be infrastructure support (roads, utilities, railroad, airport), a good communication system, and security for facilities and for workers and their families. Without the above, meaningful investment that can help improve an economy will be limited.

To relieve the economic stress, it is important to create demand for domestic products and concurrently, when possible, establish markets for sustainable export commodities as might exist (e.g., industrial, agricultural, energy, raw materials, or finished products from raw materials). The domestic path to reduce economic stress is limited because of joblessness and a lack of familial acquisition power beyond absolute essentials. Also, the inability to start small businesses because of limited loan possibilities retards local entrepreneurship and creation of job opportunities. The export path to relieve economic stress is possible because there is global need and competition for natural resources (raw materials).

Morally and ethically, international conglomerates should not purchase raw materials originating where war with ethnic underpinnings is waged to control saleable commodities (e.g., "blood" diamonds, gold). An example of this is the case in the eastern area of the Democratic Republic of Congo. Here Rwandan Hutus have joined with Congolese rebels to control mineral riches (e.g., "conflict minerals" columbite-tantalite which contain the elements niobium and tantalum that are used in cell phones). At the same time, these groups persecuted and killed Tutsis who escaped to the Democratic Republic of Congo during the 1993 genocide.

The economic stress suffered by citizens globally during the beginning of the 21st century will be exacerbated in 2035/2050 as populations expand by as much as half in many nations. There is high unemployment in less developed, developing, and developed nations alike at this writing in 2009. This is an intolerable situation for many governments and does not bode well for their growing populations. Economic stress on the populace and on nations is inexorably linked to social and political stressors.

Socio-Political Stressors

Social stressors that impact societies ensue mainly from political doctrine (principles, morals, ethics). Political restraints and governmental priorities that are not in societal interests create tensions among citizens, whatever might be their social, economic, or educational levels. This can raise socio-political pressures that have exploded in recent history and ended with changes in government, sometimes peaceful, sometimes bloody, sometimes for the good, and sometimes not. The political restraints include no free election of a country's leader or of peoples' representatives to legislative bodies, or no election at all. Restraints on free speech or right to assemble or manner to present grievances to their government compounds stress in societies. The lack of an effective and independent judicial systems and no habeas corpus provision allows infringement on human rights, failings in personal security, and little or no protection of property rights. Individually or together, these restraints intensify stress in the citizenry and arouse public restiveness.

The socio-political stress on populations is further heightened by high taxes but with little or no pass through of these finances or those from any sale of nationalized natural resources or services for the public benefit. A pass through of funds can lead to an improvement in what are inadequately supported public services (fire fighters, police, health clinics, schools), all amid a crumbling infrastructure. Social stresses are added to by a lack of educational opportunities, particularly for children and by the failure of a healthcare system to provide clinics and hospitals, fully staffed and well-stocked with medicines. Stresses are intensified by unemployment and by the failure of a government to provide the basic necessities of life discussed in a previous section (i.e., safe water, untainted and nutritious food).

Social stresses develop as well when there are religious divides such as have led to the conflict between Islamic fundamentalists who interpret the Koran as demonizing Christianity, Judaism, and other organized religions. This is not Islam true to the Koran. Fundamentalists see the only ends for followers of other religions as conversion to Islam, exile (if an option), or death, a replay of the Spanish Inquisition that survived initially but ultimately failed. In addition, ethnicity, tribalism, and clan loyalty can stress societies to the point of undeclared or declared wars (e.g., between Sunni and Shiite Arabs, between Serbs, Croats, Bosnians and Kosovoans, internecine conflicts between Palestinians and Palestinian Hamas, and between Lebanon and Hezbollah). The more than 20-year war between Tamils and Sri Lanka ended recently as the Sri Lanka military decisively defeated the Tamil rebels. Perhaps the worst of contemporary tribalism horrors was between Hutus and Tutsis in Rwanda and Burundi where the Hutus make up 90% and 85% of the populations, respectively. The Hutus wrought genocide in 1993 with the killing of between 500,000 and 800,000 Tutsis.

The socio-political issues that bring on the various classes of stress from natural growth discussed earlier can only worsen as populations increase in countries where large populations already strain national resources, some beyond their economic limits. The resources of local governments are further over-stretched by demographic changes as people move from rural areas and add to existing densely populated urban-suburban centers or create new ones.

National growth from immigration (legal and illegal) adds a troubling facet to unsettled questions about population increase. Professionals and others with skills needed by a country are welcome to immigrate whatever its population status. Welcoming republics expect that as the foreign emigrées seek a better way of life, they will integrate into a nation's economic and social fabric and become honorable and contributing naturalized citizens.

Countries with contracting numbers of citizens bring in another class of emigrées, often unskilled, as guest workers to fill service needs. Some of these nations do not encourage naturalization for guest workers and their families. They may manifest this, for example, by not offering language classes or bilingual education for children. This is a barrier against the workers and their families melding with the general society and ultimately becoming naturalized citizens of the host country. This has been the case in Japan with more than 500,000 skilled workers from Brazil who are second and third generation Japanese Brazilians. They look like the citizens of Japan, have the same values as instilled in them by their parents and grandparents, but they lack the facility in Japanese and culturally are Brazilians. As a result, the Japanese Brazilians tended to live in the same enclaves. Their children went to schools that taught in

Portuguese because Japan did not support bilingual education. With the global downturn in economies in 2008/2009, many have lost their jobs in Japan and are opting to return to Brazil even though Japan, with its vulnerability because of a contracting population, will desperately need them when its economy strengthens. Now, the country has initiated language education for adults and is supporting bilingual education in schools.[16]

Whether immigrants enter a country as professionals and skilled workers or as unskilled service providers, the newcomers maintain their cultural principles and religious identities as the bedrock of their being. They pass these on dutifully to succeeding generations. Examples of this cultural/religious continuum include Turkish guest workers in Germany and Pakistani immigrants in Great Britain.

Newcomers often seek housing in ethnic enclaves where neighbors look, think, and act as they do. They feel comfortable and have time to acclimate to the unfamiliar societies they have entered. This was true, for example, in the United Stats for Europeans who immigrated from 1840-1890 (mainly from Germany, Ireland, Sweden, Austro-Hungary, and Italy) and emigrées from Canada and Newfoundland, whether they were Catholic, Protestant, Jewish, or of another belief. It has been true as well in recent decades for Asians and Latin Americans. These groups entered into the economy of the United States, availed themselves of educational services to learn English, schooled their children, and became loyal citizens. They conformed to principal societal values and demonstrated loyalty or a "glad to be here" attitude to the country that accepted and succored them. They joined the military to defend their new country when it was threatened. This resulted in the emigrées melding with their fellow citizens in the general population while they maintained their cultural principles and religious identities.

Although most legal immigrants benefit from employment possibilities, healthcare, educational, and social/welfare services of a country, there are problems caused by small numbers of them. Those who reject their host country, its values and way of life, overtly prejudice the vast majority who have adopted their new country with its laws and the freedoms it offers. The negative stands of a relatively small number of newcomers can lead to tensions with many in the general populace. Some citizens may unjustly view all immigrants of like ethnicity or religious beliefs as the malcontents with a jaundiced eye. This becomes a major social stressor that needs relief.

This is an especially gnawing contemporary issue in countries with shrinking populations such as in Western Europe (e.g., Germany, the United Kingdom, Italy, the Netherlands, Spain, and others). It has resulted in clashes between communities that occur when inculcated differences between alienated immigrants and host populations manifest themselves in both subtle and overt, and sometimes violent, criminal, seditious, and murderous ways. These are often times driven by zealot-like religious beliefs or loathing of a way of life. As emphasized above, a grand majority of immigrants do become citizens loyal to and loving of their new country. Sadly, they may suffer injustices because of the attitudes and actions of a small number of their co-cultural and co-religionist groups.

Ecosystem Stressors From Increased Populations

There has been an extraordinary exponential increase in global population during the past century from about 1.5 billion people to more than 6.8 billion people in 2009. There has also been a dramatic move of citizens from rural settings to urban/suburban areas creating denser concentrations of populations there. The increase has been and continues to be greatest in the developing countries. As described previously, this has taxed the ability of governments to provide the basic physical, biological, economic, and socio-political needs of large segments of their societies. The demand on ecosystems worldwide to supply these basic needs and some of population "wants" has damaged ecosystems and reduced their vitality and normal productivity.

The harmful intrusions of the earth's ecosystems as a result of human activities are fostering serious concerns locally, nationally, regionally, and globally. For the most part, the intrusions are multiplying rather than diminishing. This further reduces ecosystem capability to provide sustainable resources for growing populations. Again we ask, can the Earth support a stabilized population in 2050 of 9.5 billion to 11 billion people with a satisfactory quality of life that some in 2009 have but billions if our co-inhabitants on the planet do not now have? In light of the 2009 economic, environmental, and political global situation and intrusions of ecosystems, this seems unlikely. Only rapidly decreasing fertility rates in key countries, especially in sub-Saharan Africa, but also in Asia and the Middle East, can help ease the problems associated with growing populations.

The concentrations of populations and sustainable industrial, manufacturing, and agricultural efforts will strain water supplies in many nations and draw down on other natural resources (raw materials) at a faster rate than they are being replaced. Without enforced environmental control, the activities that support the needs of growing populations will continue to pollute the atmosphere, water supplies, and soils, and hence crops and fodder grown in them. Citizens can suffer serious chronic health problems from the pollution that passes into them from breathing, drinking, and eating. The same is true for other life forms.

Physical Actions

As populations grow, housing is needed together with an infrastructure that supports them and industrial, manufacturing, and service projects, plus other programs that create employment opportunities (e.g., roads, bridges, utilities, schools). This requires land and an encroachment into environments and the habitats they contain. The encroachment means the loss of grasslands (often times agricultural terrain), forestland, and productive working farms. Forestlands are also lost to clearing in order to expand arable zones for cropping and raising livestock to feed growing local and regional populations (e.g., in the Brazilian Amazon). As natural habitats are invaded and absorbed by human communities, life forms (vegetation or animals) can be displaced or lost.

Industrial projects in the mining and energy sectors have physically re-sculptured natural topography in two ways. An example of the first is open pit mining for metals such as gold and copper, and strip mining (taking off mountain tops or leveling terrain for coal mining or tar sands extraction). An example of the second is the shaping of an altered topography from the disposal of wastes from such ventures on adjacent terrain or in pristine rivers. Both types of projects destroy habitats for resident life forms with the purpose of providing for the needs and wants of growing populations. Nonetheless, the needs and wants of ecosystems, and hence of people and other life forms can best be served by reclamation of disturbed terrain to a condition as close to the original topography and vegetation as possible. This can be mandated and enforced by law.

The physical diversion of rivers has caused biological and economic disasters for what were productive ecosystems. When the Soviet Union diverted rivers to provide irrigation for farming in the 1960s, the Aral Sea that had been fed by river flow shrank and the fishing industry that was the livelihood for large shore populations died out. Tourism to the Aral Sea shores likewise failed. Only recently, after the disintegration of the Soviet Union, have the bordering nations of Kazakhstan and Uzbekistan begun a program funded by the World Bank that is slowly restoring the Aral Sea. Similarly, the non-conservative use of water for irrigation instead of irrigating by focused drip-type water delivery systems to crops can deplete aquifer sources if the amount of water discharged is not balanced by the re-charge of water in an aquifer.

Climate change is a physical response to global warming brought on by anthropogenic activities that release carbon dioxide gas (CO_2) into the atmosphere, especially from the transportation and energy-generating sectors. These activities and increased industrialization have grown greatly during the past few generations in response to the demands of larger populations. Emissions from these sectors and others that generate substantial masses of greenhouse gases (e.g., carbon dioxide [CO_2], methane [CH_4]) must be reduced markedly, even as populations increase. Failure to do this means that the global warming caused disasters that threaten the earth's ecosystems and their life forms will become harsh realities rather than threats. Can this be achieved in the immediate future in light of the obligation of governments to provide the needs and wants of expanding populations midst the ongoing races to industrialize in many populated developing nations? At this writing (2009), the answer to this question is "unlikely to happen" in time to avert an increase in the rate of global warming and the ecosystem disruption it portends for many regions of our planet.

Biological Actions

An increased earth population will require greater food production. This is achievable if today's food growing practices are modified, if more unused arable land is put into agricultural production, and if degraded land is reclaimed for cropping. Cropping foods without nutrient replenishment with natural or man-made fertilizers has reduced agricultural productivity in many countries and left them with food deficits. The soil ecosystem needs nutrient replenishment to maintain sustainable yield of substantial and good quality farm products or an increase in both their yield and quality. Bad plowing or tilling practices for soil preparation for planting can lead to their degradation by erosion. This decreases its productivity and resulting in its eventual agricultural demise. A change in tilling practices by corn growers in the United States reduced erosion by 40%.[17]

Over harvesting of fish from the oceans to feed existing populations is a problem that the world is currently confronting. If the current rate of fish catch is maintained it is possible that 70% of marine food fish will disappear by 2035.[18] Overfishing will disrupt the food chain/food web continua in the oceans and upsets the biological diversity balance in marine ecosystems. This can conceivably drive some species toward extinction. Only controlled harvesting can assure the sustainable existence of this important source of food protein for the projected 2050 population of 9.5 billion people or more. Over hunting of animals for food will result in the same consequences.

Chemical Intrusions

Pollution is released into ecosystems daily in the form of toxic metals, toxic industrial chemicals, agricultural chemicals, gases, and particulates. Many are decreasing in concentration (e.g., sulfur dioxide [SO_2, acid rain component], lead [Pb]), while others (e.g., carbon dioxide [CO_2], nitrate [NO_3], mercury [Hg]) are increasing in concentration in ecosystem atmospheres and waters. The pollutants originate as emissions from the transportation sector, from industrial operations (e.g., electricity-generation, mining/smelting), and manufacturing plants. Pollutants derive also as runoff from agricultural fields into water bodies and onto soils and vegetation, from mines (acid mine drainage), and from improperly disposed effluent wastes from industrial, manufacturing, mining, and domestic/commercial sources. Seepage of water-borne pollutants into soils and underlying aquifers harms subsurface ecosystems. Chemical pollution undermines the pristine purity of environments and their ability to maintain productivity of natural resources that are needed to help sustain additional numbers of earth inhabitants.

Linkages

As stated at the beginning of this section, there are tight linkages among the arbitrarily designated physical, biologic, economic, and socio-political stressors. Examples are encroachment of land for housing as populations grow, a global problem, and the cutting of forests for hardwoods, and livestock raising and cropping (as in the Brazilian Amazon) to provide for growing global populations. Each creates biological stress by killing vegetation and disrupting natural reforestation. These actions severely degrade habitats. This puts stress on ecosystem inhabitants including humans to search out and move to other habitats or die slowly as ecosystems fail. Economically, the encroachment projects may yield financial benefits but these can often be short-lived. This puts socio-political stress on governments to limit the physical areas for such projects in order to preserve forest environments, and in the case of the Amazon, protect indigenous people who live from the forests and the food animals they hunt.

Similarly, diversion of rivers for agriculture or electric power to benefit growing populations will deprive fauna and flora in down flow environments of a life-sustaining resource. If the life forms or environments are a source of income for downstream businesses (e.g., capture and production of food resources, tourism), the economic raison d'être of populations will suffer. This was the situation of the Aral Sea cited previously where the sea shrank, water salinity increased, and most life in the sea ceased to exist. Ten thousand fishermen and cannery workers lost their livelihoods as did many more people in the tourism industry. These local industries were eradicated and an important protein-rich food source was lost. An undermining of economies will cause social unrest and political necessity to face the situation. In the case of the Aral Sea, the Soviet Union had to collapse into independent states before Kazakhstan and Uzbekistan, as new countries, could begin restoration of the Aral Sea with loans from the World Bank.

Another example of linkage of stressor actions can originate from cropping and animal husbandry where runoff of agricultural chemicals (fertilizers) from fields or nutrients from animal waste (from chicken farms, feed lots, slaughter houses) discharges into water bodies. This is happening today in Chesapeake Bay, southeastern United States. The nutrient-rich runoff (phosphorus, nitrate) from land stimulates growth of algae in the water to grand blooms that die, settle onto the bottoms of rivers and estuary near shore zones, stress the ecosystem by algal decomposition that uses all available oxygen causing fish and shellfish kills. As was the case in the Aral Sea, this deprives people of an important, protein-rich food source and damages economies based on fishing, harvesting of shellfish, and income from tourism. The biological and economic stresses suffered in the Chesapeake Bay (and confluent rivers) brought political pressures on the bordering states of Maryland and Virginia, and on the federal government. They passed laws with enforcement clauses that have resulted in a reduced nutrient runoff and improved economies dependent on an ecologically clean Chesapeake Bay.

AFTERWORD

Time is running out for global collaboration in actions to arrest population growth and to alleviate stress now suffered by many populations worldwide (Table 2-1). It is in the interests of citizens of all nations, whether developed, developing, or less developed to work in unison and deal with problems affecting our Earth's ecosystems. We cannot lose the productivity that provides sustenance and other products that our planet's citizenry needs to survive, exist, live well, live long, and prosper.

As we have read and learned in Chapter 1, global population is expanding. The fact is that we were not able to provide sustenance, security, and a good quality of life for 1/3 of the Earth's inhabitants when there were 6 billion people at the end of the 20th century. We are no better off in making provisions for today's 6.8 billion people. This is most evident in developing countries where citizens are continually under stress to provide for themselves and their children but where there is high unemployment and inadequate educational systems. In two generations or less, the Earth will be home to a "stable" population estimated to top out at 9.5 billion people. This figure may be under estimated by hundreds of millions or more.

The question is, how do countries that cannot provide for their citizens in 2009 lower their fertility rates and thus stem their rate of population growth? We discuss this in the next chapter.

CHAPTER 3. STEMMING GLOBAL POPULATION GROWTH - THE KEY TO ELUDE POPULATION CRISES/ CRASHES

OPTIONS

There are workable options available to save growing populations in developing and less developed nations from tenuous societal conditions and the ravages of failing ecosystems that can end in population crises or crashes. Governments and their citizens can respond responsibly, pick up the options, and deal with the biological-, chemical-, and physical-based problems that threaten their very being. Alternatively, governments with economic interests and little concern for the populace can ignorantly disavow the imminent and long-term dangers these problems pose. Adopting the latter tactic will slowly but surely bring on societal, economic, and political collapses that will cause governments to fall and entangle developed nations that will be called upon to help resolve life and death issues.

THE PRINCIPAL SOCIETAL PROBLEM

In this writer's opinion, the most pressing problem confronting civilization is population growth. The most effective approach to attack this issue is to establish educational programs that emphasize the why and how of population stability. For example, why is holding the line on population important to humans and the ecosystems that provide for them? How does a fixed population benefit citizens' social, economic and political situations? It is equally important to update existing educational programs so as to reemphasize ways to deal with the population growth question and its consequences if they are not applied and subsequently provide viable solutions to it. This chapter will discuss how citizens can justify a moderation of their religious, cultural, and ethnic norms with the end of scaling back on fertility rates and hence population growth rates.

Population growth is a prime driver of many linked ills that threaten social equilibrium in much of the world. Thus, when we consider the subject of growing numbers of people on earth we must contemplate how to address these other ills in order to improve the quality of life for all earth inhabitants. Fundamental questions for which we must seek workable answers include the following: 1) how to alleviate physiological and psychological stress from lack of sufficient nourishment in populations by restructuring the management of water resources and food production and their distribution practices; 2) how to attack pollution problems locally as well as regionally and globally in order to protect people and the ecosystems that sustain them; 3) how to efficiently and effectively reform healthcare and educational delivery programs; and 4) how to improve economies by attracting investment that will create employment opportunities for educationally prepared persons as well as those with limited education. These are basic societal issues that must be dealt with by governing bodies for the benefit of their citizens and thus secure their families futures. With realistic thinking, creativity, technological advances, and the will to find answers to the questions cited above, these issues and others can be solved individually and then blended into a master program that will set the foundation for stable, healthy, and prosperous communities. We will discuss these and other societal problems in light of population status in subsequent chapters.

POPULATION STABILIZATION

During the decade of the 1950s the global population increased from 2.52 billion to 3.02 billion people in 1960 for an average annual change of 50 million people. With a few exceptions (1958-1960, 1974-1977) the rate of annual population growth increased until it peaked in 1989 with more than 88 million people added to the global inventory. The global population then was about 5.2 billion people. Annual population growth decreased until 2002 with slightly more than 76 million people added to bring the earth populations to 6.25 billion people. Since then the annual addition has slowly crept up and is projected to reach about 80.5 billion in 2012 when more than 7 billion people will inhabit the earth. The most recent statistical projection (2009) is that the annual population growth increase will fall steadily after 2012 with about 48 million added in 2049 giving a 2050 global population of 9.5 billion people. Only a few years before, the projection had the 2050 peak population at 9.2 billion people. Clearly, these population estimates are subjectively objective figures.

If, as some have proposed, the earth has a carrying capacity of 7-7.5 billion people without major population crises (from sickness, famine and starvation, war), this range will be reached between 2012 and 2018. It is a coincidence that the 5126 year old Mayan calendar sets December 21 (or 23), 2012 as the date for a major event to affect the Earth. Could this be something positive like the discovery of a vaccine against HIV/AIDS or one against dengue fever? Could it be the discovery of an economically feasible method to remove carbon dioxide (CO_2) from the atmosphere to its level in 1990, or ideally to pre-global warming concentrations? Can the captured gas be sequestered or efficiently converted into carbon and oxygen? Could the event be a spiritual awakening that causes all peoples to respect other religious beliefs and cultural norms, resolve political problems, and bring global peace and prosperity? Or could it be something negative such as a tipping point time for population crashes, the date of an asteroid strike, the day for the beginning of a series of devastating global volcanic eruptions, or the time of the beginning of worldwide killer pandemic? Or could it simply be a benign date that marks the yearly solstice? In a few years we will know.

METHODS TO LOWER FERTILITY RATES AND REDUCE POPULATION GROWTH

There are several ways to slow and arrest population expansion before the numbers of people versus sustenance/health limitations evolve into a population crash scenario. One plan to reduce fertility rate is mandated by national law. Others related to education and economics have proven to be effective. Impediments to global programs that aim to reduce population expansion are founded on religious doctrine, cultural mores, and ethnic practices. In light of what could become a tragedy for mankind that will impact people mainly in less developed and developing populations, wise and honorable persons are working to moderate the traditional positions in this religious, cultural, ethnic grouping.

Government Mandated

The People's Republic of China (PRC) has about 20% (1.35 billion people) of the world population which it has been able to feed with 7% of the world's arable land. During 1979, the socialist/communist government realized that population growth had to be controlled to alleviate future social, economic, and environmental problems. A law was enacted that permitted one child per family under the penalty of fines, a move to less desirable housing, or job demotion. There was a caveat that a family could have a second child if the first was born with physical or mental problems. Exceptions to the law are made for rural couples that need children to work the fields and care for livestock, and for ethnic minorities (2 or 3 children). This means that 35.9% of the population, mainly in urban centers, is subject to the legal restrictions whereas 64.1% of the population, mainly in rural areas is not. However, there was a recent exception made for Shanghai, a city of 16 million people that has an aging population with more than 21% of its citizens older than 60. Here, the government allows couples that are single children themselves to have two children. The law has been effective if PRC census figures are accurate. In mid-2009, the PRC rate of population increase was 0.5% with a fertility rate of 1.6, well below the 2.06 replacement rate. Clearly, such a mandate would not be tolerated in a democratic society.

India has a law that allows only couples with two or fewer children to be eligible for election to local governments at the village level (>500 people). India, with a mid-2009 population of 1.17 billion people has a rate of natural increase of 1.6% and a total fertility rate of 2.7, well above the replacement rate. This can lead to a doubled population of >2.3 billion people by 2050 if rates do not continue to decline. The Chinese population is in a contraction mode while that of India is in an expansion mode that threatens the future social and environmental fabric of that country even as its economy expands. Only about 56% of the women of childbearing age in India use contraceptive methods versus 90% in China.[19] Education programs and incentives to increase the use of contraceptive methods to reduce the Indian fertility rate are essential for India to move towards population stabilization.

Education
Education has two paths towards reducing population growth. One applied in Iran is legal and perhaps would be

better placed in the previous section. Iran has a law mandating that contraceptive courses are required for males and females before they can obtain a marriage license. Although Iran is governed by a conservative religious regime, Persian sagacity in the pragmatic ruling clergy deems a stable population as one major factor necessary to obviate future socio-political, economic, and environmental issues. This has worked for the country of 70 million people. The Iranian mid-2009 natural growth rate is 1.5% and the 2.0 fertility rate is at the replacement level.

Surely, family planning clinics in nations with high natural growth rates are a medium through which to make both females and males aware of the social and economic implications of having several children and the benefits of having fewer offspring. For example, one truism the clinics emphasize is that family funds spent on a smaller number of children can lead to their better quality of life (e.g., nutrition, healthcare, education). Conversely, too many children to support dilute family funds and make basic needs less accessible to them. Education can be the gateway to employment and its psychological and financial rewards, and social freedom. Additionally, a woman with fewer children to care for can seek education or an apprenticeship that would prepare her to work (part- or full-time) and contribute to a family's economic well-being.

The second path is schooling for both boys and girls and for adults if they are so inclined. Public schooling in less developed and many developing countries may not be available to all because of government financial constraints. Also, when available, schooling may be too costly so that only a relatively small number of students from financially advantaged or politically connected families attend. This is true in several countries in

Africa and some regions of Southeast Asia, South America, and the Middle East.

A barrier to education for girls and women is found in some Islamic Fundamentalist groups such as the Taliban in areas under their control. At these areas in Afghanistan, the Taliban prohibit girls from going to school, bomb or burn down schools, and maim or kill teachers who attempt to educate females. Yet educated women tend to have significantly lower fertility rates than uneducated or poorly educated women in the same country. Educated women can work outside the home as well as caring for their families. The outside employment contributes to a family's income and access to products that they previously had to save for a long time to purchase. The empowerment of women as a result of education and earning power is sensed as a threat by some male-dominated societies.

Religion and Culture

Religious doctrine or cultural traditions that lead to larger families or does not discourage them counters strategies designed to reduce fertility rates to replacement levels and a move to towards global population stabilization. The Catholic Church with 1.1 billion adherents worldwide is doctrinaire and prohibits the use of modern contraceptive methods. The Church recommends abstinence or following a menstrual calendar to limit the possibility of pregnancies.

Islam, with an estimated 1.5 billion adherents globally, encourages large families as a tradition of the Islamic culture but does not prohibit the use of modern contraceptive methods. Nonetheless, many Islamic-dominated countries do tend to have high fertility rates and hence higher natural population growth rates. Other countries representing more than half

the world Islamic population are aware of the future implications of growing populations and do not discourage the use of
contraceptive methods. The latter group has worked through
education and incentives to bring fertility rates down and thus
reduce their natural growth rates. Table 3-1 lists examples from
both groups.

Table 3-1. Fertility rates and natural growth rates of selected
Islamic-dominated countries.[19]

Fertility Country	Natural Rate	% Growth Rate
Iran	2.0	1.5
Turkey	2.1	1.2
Indonesia	2.5	1.5
Malaysia	2.6	1.6
Bangladesh	2.5	1.6
Egypt	3.0	1.9
Sudan	4.5	2.2
Pakistan	4.0	2.3
Iraq	4.4	2.3
Syria	3.3	2.5
Nigeria	5.7	2.6
Afghanistan	5.7	2.1
Saudi Arabia	3.9	2.6
Somalia	5.5	3.0
Yemen	6.2	3.0

In the near future, we will see increased suffering of an
expanding poverty stricken global population. As abject poverty worsens because the carrying capacity of ecosystem Earth
is strained to provide for an increased earth population, one

would hope that religious leaders and others who oppose using modern contraceptive methods to reduce population growth will recognize the urgent need to do so. Respected leaders have the influence to moderate religious doctrinaire barriers or to issue official edicts that loosen the norms of cultural traditions. In these ways they can promote the efforts towards bringing down the global fertility rate to a 2.06 replacement rate or close to it.

In some cultures, especially in rural areas, children supply needed labor to work on farms and are counted on to provide for parents when they reach old age. Thus, more children bring more help and more social security. It is important in some cultures to assure that the family name is carried on. Thus, children are conceived until there is a male heir and may continue driven by the concept that more male heirs provide greater assurance that a family name will be carried by future generations. As above, knowledgable leaders in communities guided by culture and tradition can work with the populace and modify parents' desires to have large families. They may emphasize two possible scenarios. First is that as offspring marry and divide land as families grow, smaller and smaller plots may not be able to support a family's needs much less provide for their parents security in old age. Second is that income expended on a smaller number of offspring can give them a better education that can lead to job opportunities that reward them financially and provide security for aged parents.

However daunting and unnerving it may be, reduction of fertility and holding the line on the world population is necessary if we are to avoid population crashes. As we have emphasized previously, 1/3 of the earth's populations is in a day-to-day fight for survival. More than one billion people are under-nourished and suffer the symptoms of malnutrition. More than one billion are without safe water. About 1/3 of the global population have half the water they need. More than two and a half billion are without acceptable sanitation. Hundreds of millions of today's global citizens are chronically sick and millions die of diseases annually. Problems like this will be accentuated in larger populations and most damaging for the socially, economically, and politically disadvantaged citizens in less developed and developing nations. They will impact societies, economies, and governments in developed nations as well. Leadership that can cope with religious beliefs, different cultures, and fundamental traditions to establish a sustainable world population will be under great stress. Some will be unable to take the necessary decisions because of their deep inner beliefs. Others will have to stand in their stead and work within the reality of survival and prosperity or they will surely always be trying to manage survival. Tough choices have to be made. Dedicated and strong leaders with a clear sense of the 2009/2010 socio-political and economic realities and those projected for the future will make these choices.

AFTERWORD

It will take a few generations or longer before many nations with higher than replacement fertility rates will see them decrease and approach the 2.06 replacement level. Until then, populations will grow and nations will have to provide the sustenance that nourishes their very being.

Principal among the sustenance needs is water. Without water, humans can survive about 4 to 6 days before their organs shut down and they die. Without sufficient water for irrigation crops will render lesser yields of poorer quality foods. Over time, this will result in under-nutrition for consumers. Food animals wither away and die without sufficient water as happened with cattle herds in Argentina during 2009. The loss of livestock further depletes a nation's food supply and export earnings. Without water for sanitation, disease can ravage populations especially if they have poorly functioning health systems. This is the situation in Zimbabwe where cholera has killed more than 4000 people and has sickened almost 100,000 more from December 2008 through the second quarter of 2009.

With water but without food, humans can live for about 6 weeks on their body fats before they die. Without sufficient foods of good quality, populations will suffer malnutrition. Within a malnourished population, children up to 9 years old will suffer most with stunted growth and impaired cognitive functions. This will limit how they can contribute to a nation's socio-economic development.

These essences of life, water and food, are scarce in regions of many less developed and developing countries with growing populations. The next chapter focuses on what can be done in these countries to increase the safe water and untainted nutritious food supplies for their citizens now and in the foreseeable future.

CHAPTER 4. SUSTENANCE FOR A STABLE GLOBAL POPULATION FIND MORE WATER + GROW MORE FOOD: FOR HOW MANY?

Water and food sustain life. Without this nourishment life weakens and dies. Water to drink, water for hygiene, water for sanitation purposes, and water for irrigating crops and animal husbandry is essential to health. Water is essential to provide a food supply for human populations. Water is necessary to sustain ecosystems' habitats that nurture all life forms and preserve natural resources on which humans depend.

WATER

There is enough water on the planet to service the needs of the existing Earth population of 6.8 billion people. However, the reality is that these needs are not being met. This is the case because surface and subsurface (aquifer) water sources are irregularly distributed geographically. Developed countries have built infrastructures to move water from where it is in excess to where it is in deficit so that there is enough water to meet the needs of their populations. In several less developed and developing countries where many populated areas suffer from a lack of water, there is no infrastructure to move it from where it is plentiful to where it is scarce. Water problems will intensify in many countries, whether developed, developing, or less developed, because of growing populations and happenings such as global warming and climate changes that will disrupt water flow patterns and reduce drinking water availability (Table 4-1). The Intergovernmental Panel on Climate Change predicts that 2 billion people will not have access to clean water by 2050 with the figure rising to 3.2 billion by 2080, tripling the situation in 2010.[20]

Although there may be more water on earth in the future to feed the hydrologic cycle because of melting glaciers and more moisture in the air to feed storms, water sources fed by seasonal melting of glaciers will be reduced in all parts of the world (e.g., Europe, the Indian subcontinent, South America, the United States, China). In some regions, water sources will literally dry up as glaciers continue to recede because of global warming and ensuing climate change and may ultimately disappear. This will deprive large numbers of populations of their principal sources of water for drinking, cooking, personal hygiene, sanitation, crop irrigation, and animal husbandry.

Uneven distribution of water and its availability nationally, regionally, and globally will continue to be the norm. Locations with diminished water resources may become more water sufficient as a result of changing climates while others with sufficient water now may develop water deficits later. Changing and uneven area distribution of water sources will be a more severe world problem in the future than it is today if global warming and the resulting climate changes continue unabated. We will discuss this in a subsequent chapter.

Table 4-1. Population growth will result in reduced renewable freshwater supplies per capita for selected countries in the Middle East, Africa and Asia. The changes take place in a generation from 2007 to 2030. Water stress is grave with <1000m³ per capita and marginal with >1000m³ but <2000m³ per capita.[19]

Country	Population Mid-2007 x10⁶	Per Capita Water 2007 m³	Population 2030 x10⁶	Per Capita Water 2030 m³
Water Stress Intensified (Grave)				
Algeria	33.8	423	44.7	320
Burkina Faso	14.7	890	26.5	494
Burundi	8.5	442	17.2	218
Egypt	75.5	759	104.0	551
Israel	6.9	240	9.2	180
Jordan	5.9	148	8.5	103
Kenya	37.5	839	62.7	502
Libya	6.1	99	8.4	72
Morocco	31.2	895	39.2	712
Rwanda	9.7	551	16.6	393
Saudi Arabia	24.7	93	37.3	62
Tunisia	10.3	442	12.5	364
Yemen	22.3	184	40.7	101
Water Scarce to Water Stressed (Marginal)				
Ethiopia	83.0	1355	137.0	821
India	1169.0	1670	1407.7	1297
Iran	71.2	1931	91.1	1509
Somalia	8.7	1620	15.1	933
From Water Sufficient to Water Scarce				
Afghanistan	27.1	2015	53.2	1026
Ghana	23.5	2314	34.2	1590
Nigeria	148.0	2085	226.8	1361
Tanzania	40.4	2291	65.5	1413
Uganda	30.9	2133	61.5	1072

Improve Water Availability Find More Water

The obvious approach towards improving a local water supply is to have hydrogeologists locate likely areas for aquifers and drill there. In Africa, for example, hydrogeologists estimate that less than 5% of aquifers have been found and tapped. With good fortune, aquifers that contain safe quality water will be found where water is in deficit and wells will be developed to extract the commodity. Some aquifers contain fossil water that once removed is not renewed. Continued water withdrawal without recharge will empty the aquifer. If the aquifer is one that is recharged, a balance between the volume of water discharged and the volume of water recharged assures a reliable water supply for municipal wells and wells that supply farms or individual homes. In all cases, water discharge has to be managed astutely so as not to lower the water table excessively between times of pumping and times of recharge. The volume yield will determine the prioritization for its use, whether for domestic needs, commercial use, agriculture, or industrial and manufacturing use.

Although well water may test as being of good quality when an aquifer is first pumped, physical/chemical conditions in the subsurface can change because of seasonal lowering of a water table as a result of pumpage. This lowering can expose aquifer rock to air that causes oxidation and results in chemical reactions as the aquifer recharges and the water table rises and water reacts with the minerals in the exposed rock. These chemical reactions can release contaminants during the following seasonal pumpage. This happened in India and Bangladesh where arsenic (As) was released into aquifer water, poisoned large numbers of farmers and their families, and threatens millions more. Thus, the aquifer water must be tested on a regular schedule to assure that the quality remains good for the user/consumer. If there is a pollution problem, the pollutant(s) may be subject to removal at the wellhead. When this is not feasible, the water can be moved to a treatment facility before it is made available for use.

Desalination of Ocean Water and Other Saline Waters

Population centers that are deficient in safe water supply and are located in marine coastal zones can use desalination of ocean or seawater as a source of safe water. Desalination is an expensive, energy consuming process but one that can deliver potable water. Saline or brackish inland water bodies and saline or brackish aquifer water can also provide the feedstock for desalination plants. In the cases of aquifers, feedstock water has to recharge at a rate equal to the volume of safe water produced to assure a long-term desalinated water supply.

The two main desalination methods are multistage flash distillation and reverse osmosis. The distillation technique heats seawater at pressures less than atmospheric pressure so that the seawater boils at <160°F. At <160°F scaling (buildup of foulant salts) in the system is greatly reduced. The system is programmed so that the seawater passes through progressively lower pressure conditions where the steam flashes off at steadily decreasing boiling temperatures and is captured. This is an energy intensive technique. At present 85% of the global production of potable water (>12 billion gallons of water per day) is generated by the flash distillation method. Seventy-five percent of this production is in the Middle East.

In the reverse osmosis technique, seawater (or gray water from domestic, commercial, or industrial/manufacturing waste) is pressure driven through semi-permeable membranes that separate salts from water. The pressure ranges from 800 to 1200 PSI for seawater and from 250 to 400 PSI for brackish water depending on the specifications of the membranes used. Today's membranes are made from polymers of polyamide plastics. They are manufactured specific to the characteristics of the seawater or wastewater feed (e.g., salinity, and contents of particles, chemicals, and organic matter). For some projects, dissolved matter is separated from the water by sequential passes through membranes with decreasing size characteristics to better purify the final product. In some facilities, the final stage in water purification is a pressure driven pass through nanofilters that can capture bacteria, viruses, and pesticides. Reverse osmosis uses less energy than thermal distillation and is the fastest growing desalination technique. Of the 13000 to 14000 desalination plants presently operating worldwide, 50% are based on reverse osmosis systems. Improvements that can make the reverse osmosis technique several times more efficient are being investigated at industry laboratories and at universities often supported by government funding. The research includes designing technologies use of solar energy and geothermal energy in desalination projects.

Whichever desalination method is used, whether through the technologies functioning now or those that are being researched and will be used in the future, there is the question of how to safely dispose of the separated salts. Every cubic meter of ocean water will yield about 77 pounds of salts. Simply dumping the extracted salts back into the water body from which they came may not be the answer to safe and secure disposal of this industrial waste. Conceivably, the salts can be sold to a chemical company that can process them to extract elements and compounds that can be used by society. For example, the products could be purified salt (NaCl = sodium chloride), the condiment, salt used to melt ice and snow from roadways and sidewalks, potassium (K) for use in fertilizer or other compounds, chlorine (Cl) to make bleach or hydrochloric acid (HCl), sulfate (SO_4) to make sulfuric acid (H_2SO_4), bromine (Br) for pharmaceuticals, and others. The planning to know how extracted salts will be disposed of or used should be completed and approved before a desalination facility goes into operation.

Treat and Recycle Wastewater to Deliver Safe Water

Populations in developed countries receive most of their safe water supplies from treatment plants. The plants collect and treat wastewater from domestic sources (mainly sewage), commercial sources, and effluents from industrial and manufacturing factories. They also purify tainted waters from rivers and other sources and deliver safe water to consumers through piped distribution systems. They recycle the clean waters into the domestic water supply or discharge them into surface water sources where they become part of the hydrological cycle. Safe waters also come from aquifers where they are withdrawn at municipal wells, farm wells in rural areas, and at individual household wells.

Less developed and developing countries where there are wastewater or contaminated surface waters that could be made safe often do not have purification capabilities to serve all their citizens. With treatment facilities, safe water supplies for populations can be increased. However, many governments do not have the economic capability or the priority to invest in the construction of an infrastructure to collect and treat wastewater, distribute safe water throughout their countries, and maintain the clean water plants. The failure in prioritization is one cause of widespread chronic diarrhea sickness in the non-serviced populations and can result in epidemics such the cholera problem in Zimbabwe discussed in Chapter 2.

As populations continue to grow in less developed and developing nations, investment will have to be made to provide more safe water to supply greater numbers of citizens. This means building collection and treatment facilities that use tainted river water, industrial effluents, and sewage as feedstock, and systems for the distribution of cleansed water. It means maintaining the total system so as not to interrupt the continuum of collect, treat, and deliver stages. This can prevent sicknesses in populations and avoid epidemics from sanitation failures as happened in Zimbabwe during 2008/2009.

The treatment of wastewater, whatever its origin, takes place in two or three stages depending on the types of contaminants that have to be removed. The primary treatment removes solids (raw sewage), first using screens, second using grinders to comminute solids that pass screens to 0.03 cm in diameter, and third passing the separated fluid in a slow flow into sedimentation tanks. During this phase any grit and sand still present settle out while fine-size organic matter flows through. Solid matter that clogs the screens is automatically cleaned off. The secondary stage decomposes the remaining organic matter by oxidizing it in an aeration tank with help from aerobic microorganisms, thus reducing the biological oxygen demand (BOD). This stage gives carbon dioxide and water as products and more microorganisms that are reused. The tertiary stage removes phosphorus, nitrogen, and other nutrients or toxic substances that remain in the water after the primary and secondary treatments. This stage uses reverse osmosis and biological and/or chemical reactions that precipitate undesirable components from the water.

Sludge accumulated at the collection and treatment plants has to be used or disposed of to prevent it from reentering and contaminating sensitive ecosystems. Sewage sludge is organic and nutrient rich. If chemical analyses show it to be free of inorganic pollutants such as heavy metals, it can be treated to eliminate bacteria and viruses and dried for use as a fertilizer. Depending on the organic matter content, dried sludge may be used as feedstock for combustion in industrial processes. Creative planners can devise functional options for the disposal of such wastes.

Nanofiltration: A River Water Purification Technique

Water from the Val d'Oise river, France, is fed to the Méry-sur-Oise water treatment plant. The polluted river water is decanted and chemically treated for a few days to allow settling of solids and precipitates. Next it is subjected to ozonization, then filtered through sand, and finally through charcoal to further cleanse the water. The clarified water is then pressure driven at 8-15 bars through spiraling tubes of nanofilters presenting more than 3 million ft^2 of filter surface. The manufacturing process tailors the nanofilters according to the characteristics of the feed water. At the Méry-sur-Oise plant, the filters remove bacteria, viruses, and pesticides down to one 10,000th of the thickness of human hair (0.001 micron range). The process does not remove heavy metals. Engineers avoid the problem of fouling deposits of biologic matter of microbial origin at the nanofilter surfaces and of scaling by inorganic foulants by incorporating an automated system using anti-scalants and other cleaning agents. The production capacity of this treatment plant that services more than 300,000 households north of Paris with safe water is 140,000 m^3/day (37 million gal/day). Capital investment for this plant was less than 200,000 euros and the cost of production is about 0.10 euros (US$0.15) more than that of a conventional treatment plant.[21, 22]

Import Water

Another approach to augment water supply is to import water from where it is abundant to locations where it is deficient. This is done via pipes perhaps with pumping assistance and via canals or aqueducts using gravity feed. The Romans used aqueducts to supply water to Rome and to move water in other areas of their empire (e.g., Spain, the Middle East). New York City, with 9 million people, imports water via tunnels from the Catskill Mountains watershed 125 miles to the north. Los Angeles, California, a city of 3.9 million inhabitants and a metropolitan population of 14 million people bring water through a system of canals from 400 miles away in northern California and from the Colorado River 300 miles to the east. Other population centers and agricultural operations in southwest United States do the same and by law are going to receive a significant portion of what is being imported by California from the Colorado River. To make up for the deficit and to satisfy the needs of its growing population, California is planning for the construction of a series of desalinization plants along its southern coast to cope with the loss of a significant volume of its Colorado River source.

Water From Air

Humidity can be extracted from the atmosphere to provide a new source of water for drinking, cooking, and personal hygiene. An Israeli company worked on an original technology based on United States patents and developed and enhanced them to manufacture a water extraction and delivery system. Basically, air is moved through filters to remove, particles, pollutants and microorganisms, and then to a desiccant (silica-based gel granules) that absorbs humidity spontaneously. The water desorbs from the desiccant as steam by wind-drying, vacuum, and moderate heating. The steam condenses spontaneously at relatively low temperature to yield clean water. Cooling speeds up the condensation process and preserves heat energy that is recycled into the system. Minimal electrical energy is used in the extraction to condensation phases.[23, 24]

The company manufactures equipment to meet domestic needs and machines that can yield up to 1000 m^3 daily. One of the primary aims of the development of this extraction equipment is to provide clean water for third world villages where wells have dried up or where well waters are contaminated and villagers have to walk kilometers to find safe water, or where piped in water is of poor quality. In a recent test in Jalimudi village, (Andhra Pradesh state) in India's southeastern hinterlands far from pipelines, the extraction of water from air equipment provided 5m^3 daily (5000 liters) for 600 villagers to use for drinking, cooking, and washing. The Indian government met the electrical needs.

Ongoing research is evaluating the use of solar- or wind-generated electricity to power the equipment in out of the way villages where the climatic conditions favor the use of sun and wind. The $100,000 cost for this small but sufficient water yield equipment for small villages (600 people) is feasible with government or international institutional help. Where water is the premium commodity, politics are set aside. The Jerusalem-manufactured machines have found their way to several Middle East nations (Saudi Arabia, Kuwait, Iraq, Jordan, Egypt, Qatar, and Oman). They are also operational for the Chinese Navy, South African Army, U.S. Marines, and for hospitals in Bolivia and Venezuela where the tap water is contaminated.

Conservation/Economics/Collaboration

Certainly, conservation of clean water will extend the supply and make more available for people, farming, and industry.

This means investing in simple but effective programs such as use of low volume flush toilets, shower heads that minimize flow, repair of dripping faucets, drip irrigation in farming, and recycling systems in industrial/manufacturing facilities. Pricing water with fees that are commensurate with purpose and volume used assures that the economic impact is fair to all citizens. Fair pricing for normal water use and higher pricing for overuse of this precious commodity can bring about the conservative use of a water supply.

All of the methods and programs discussed above, and others that are used (e.g., electro-dialysis, solar and geothermal energy based) are designed to find or generate more safe water and to preserve existing water supplies. Collaborative projects acting locally, nationally, and regionally can improve the supply of clean water for existing populations and for future generations.

FOOD/NUTRITION

Food is the fuel, the nutrition that energizes human, animal, and vegetation activity on our ecosystem Earth. Small-scale farmers who have supplied foods for local populations are no longer able to produce because the prices for seeds, fertilizer, and other farming needs have doubled since 2006. This leaves a void in terms of food security. Although there is enough food produced globally to feed the 2009 population of 6.8 billion people, it is not getting to more than 15% of the world population. This represents more than one billion persons, mainly children, who need it to stave off the ravages of undernourishment and starvation. Most of this population live in developing nations with 2/3 in seven countries: India, China, Congo (Kinshasa), Bangladesh, Indonesia, Pakistan, and Ethiopia.[25] Undernourishment (malnutrition) is the condition that develops when the body does not get the right amount of vitamins, minerals and other nutrients it needs to maintain healthy tissue and organ function (i.e., sound physical health, normal growth, and typical emotional development). The nutrient deficiency condition is pronounced in developing countries with 17% of the people malnourished compared to <2.5% in developed countries. It is greatest in sub-Saharan Africa where 30% of the population suffers from malnutrition, followed by Asia (excluding the Middle east) with 15% of the population being undernourished.[25] Chronic malnutrition manifests itself by stunted growth, reduced cognitive ability, and greater susceptibility to disease. The stunting is not reversible and inability to learn at a high level hurts a country's capability of having an educated work force to contribute to its economic development. During the mid-1990s, more than one million people are estimated to have died from starvation in North Korea and millions more suffered from under-nourishment. The result is that North Korean teenagers are 4 inches (10+ cm) shorter and 25 (11+ kg) pounds lighter than their South Korean counterparts and demonstrate a lesser degree of academic achievement. This attests to severe under-nourishment that reflects the scarcity of food and lower nutritional value of the foods the population ate. Under-nourishment is prevalent in many areas where population growth is the highest. Without doubt, the food supply has to be increased or supplemented where there are deficiencies if the citizenry is to be healthy and productive. The food situation can be improved in many countries where it is in deficit. Delivery of comestibles can be guaranteed where they are deficient assuming cooperation from governments, their security forces, and the international community.

The food supply and its availability where an external supply cannot be increased can be improved in three ways. First, additional arable land can be put into production where water from rainfall and/or irrigation is sufficient to support farming. This precludes cutting down forests to create land for farming or animal husbandry as has been and still is the situation in parts of the Amazon River basin (e.g., Brazil, Peru, Ecuador). Razing of forested land is counterproductive to sustaining the vitality and viability of our Earth's fragile ecosystems. It enhances the threats to humankind posed by global warming and climate change. Second is the development of food crop species that are disease, drought, and pest/herbicide resistant, give higher yields, and have greater nutritional value. Third, international agreements by all governments that guarantee free passage for the direct delivery of food to populations that can receive it when it is needed is absolutely essential. Delays in this humanitarian effort are unacceptable.

Increase Global Food Supply

There are several ways to increase food supplies worldwide. They are biological, physical, chemical, and political. Some are natural and others are artificial and require manipulating nature. Some are applied now in many countries. Others are not being used for different reasons ranging from economic, to cultural, to habit, and to concerns about their effects on human health and ecosystems. Some have been used for centuries while others are recent in their development and application. We will begin our discussion with hybridization, a natural process that has been used for centuries to breed new varieties of plants for aesthetic reasons and practical ones such as improving agricultural product yields and quality.

Hybridization

Conventional hybridization is a slow process. First, it involves finding a plant from the same species (intraspecies) with one or more sought for desirable traits. Second involves the breeding of one or more of these properties into a cultivar species to increase food stocks. Traits are sought that improve resistance to disease, pests/herbicides, and drought, and to increase yield (per acre/hectare) and quality (nutritional value) of crops. When successful, the hybridized intraspecies form carrying the desired attribute(s) will provide the parent for developing an improved seed stock. Clearly, successful hybridization serves humankind well by increasing food availability and food quality for today's populations. The expectation is that it would do the same for greatly increased populations in the one or two generation future. However, traditional hybridization is a slow process that in recent years has benefited increasingly from marker-assisted selection (MAS). This advance has significantly sped up the conventional hybridization process to improve commercial cultivars. It is discussed in the next section. The hybridization process does not use molecular or gene-splicing as does genetic engineering. In contrast to genetic engineering (gene manipulation/modification [GM]), the hybridization process does not use interspecies individuals or molecular or gene splicing to improve the yield and quality of food crops or crop resistance to disease, pests/herbicides, and drought.

Hybridization Using Marker-Assisted Selection (MAS)

Deoxyribonucleic acid (DNA) transmits genetic information. The DNA is packaged into chromosomes that are located within the nucleus of every cell in an organism. It contains all of the chromosomes that collectively make up the genome of that organism.

Marker-assisted selection is a process in which a marker based on DNA variation is used for the indirect identification and subsequent selection of a genetic determinant. The determinants sought are properties such as cited previously for food crops. Marker-assisted selection gives a time boost to standard hybridization by using DNA data from a specimen at the seedling stage to identify desired properties. This cuts the time by about half compared to conventional hybridization to find individuals within a cultivar species or wild species with sought for genetic qualities. The selected intraspecies individuals can be cross-bred with commercial varieties using traditional hybridization to develop the next generation of cultivated crops.[26] Thus, the same species with improved properties will be the parents for a new generation of food crop seed stocks. The process can be affected by environmental parameters and expected progeny differences.

Genetic Engineering = Genetically Manipulated/Modified Forms

The life, growth, and unique characteristics of organisms depend on its DNA. Genes are segments of DNA that are associated with specific traits or functions and many can be identified. The process of genetic engineering involves genetic manipulation and modification (GM) in which a gene that carries a desirable property in one species is cut from it and is spliced into the DNA of another species that lacks the trait. This creates a genetically modified organism (GMO). Scientists from biotechnology companies involved in GM (e.g., Monsanto, Bayer, Syngenta) believe that this is an efficient and inexpensive way for agriculture to increase the global food supply. Their efforts in interspecies gene splicing are directed to the same ends as hybridization: increasing crop yield (per acre/hectare) and quality (nutritional value), and enhancing crop resistance to disease, pests/herbicides, and drought. Thus far, much of the genetic engineering focus has been on improving yields and nutritional values in staples such as corn (maize), rice, soy, and wheat for populations in less developed and developing countries.

Conservative bio-scientists and health professionals do not favor interspecies gene splicing because of still unanswered questions about the effects of consumption of transgenic foods. Are they safe for humans to eat short-term or long-term or will they cause sickness? How can transgenic feed crops affect animal products (e.g., cattle for beef and dairy products, chickens for eggs) and humans who consume them? Can the ingestion of GM-food products over short or long periods of time cause gene mutations in consumers? Can crops grown with GM seeds be restricted to the fields where they are sowed or can they spread to nearby agricultural fields by wind-blown or bird/animal-borne seeds and then interbreed with natural crops causing problems? The latter situation was observed for rice crops. What might be the effects on medical patients of pharmaceuticals manufactured using GM plants? Because of these and other very real concerns, and still failing answers to them, health ministers in the European Union do not allow the entry of GMO products into their countries. On the basis of experience with GM crops and products manufactured with them, the EU concerns are well founded. There is no question that there have been human and animal health problems associated with the ingestion of GM crop prepared food or feed and also after contact with or respiration of airborne pollen during the flowering at GM maize fields. Experimental work with protein transfer between unlike species has raised some red flags about whether dangerous physiological responses that affect rats fed on GM feed can affect humans in a similar or the same way.

For example, animal feed GM-corn that was erroneously delivered to a franchised food chain in the United States was used to make taco shells. Consumers of the tacos suffered stomach sicknesses. Government investigators and epidemiologists found the tacos to be the cause. Products made with the GM corn stock were recalled and destroyed. The food chain won a legal case against the GM producer and received compensation for lost business in all of the franchises and for customers whose health had been compromised.

In the Philippines, fields planted during July 2003 with Bt maize seed (Dekalb818YG with Cry!Ab from the soil bacterium _Bacillus thuringiensis_) came into flower. Villagers living within 100 m of the maize fields suffered from fever, respiratory distress, and intestinal and skin problems. The onslaught of the sickness corresponded with the maize flowering period. Ninety-six people became sick and 5 died. Also, 37 chickens, 9 horses and 4 water buffalo died after feeding on the GM maize. Blood samples taken 4 months later from 38 villagers who still had symptoms showed antibodies to the Bt toxin Cry!Ab expressed in the GM maize. A similar situation occurred farther south in the Philippines during July 2004 with an apparent epidemiological relation between the flowering of the Bt maize and 32 villagers suffering from head- and stomachaches, dizziness, diarrhea, vomiting, and difficulty breathing.[27]

In laboratory research on the health safety theme, when a previously harmless protein in beans was gene-spliced into peas, it caused lung inflammation in mice and evoked reaction to other proteins in the pea feed. Immunological and biochemical research on the transgenic protein in the peas showed that it processed differently in the alien species. The transgenic protein from the beans altered a harmless protein in the peas into a strong transgenic protein immunogen that provoked dangerous food allergies. This raises the question of whether such a gene splicing will affect humans who consume transgenic foods and potentially be a threat to public health. The point is that the immunogen potential of all transgenic proteins should be thoroughly assessed before transgenic forms are allowed open access into the food product chain.[28]

Today, citizens in many parts of the world, with the European Union countries excluded, consume transgenic food products such as rice, tofu (soy), beef, and milk. Not enough time has passed for epidemiologists to identify any long-term physiological effects these may have on humans. Thus, we do not have sufficient data to evaluate whether consumption of GM products can cause a gene mutation in humans. This must be established before the wholesale entry of transgenic food into the food chain. Otherwise, consumers are at risk.

An additional note is necessary. Genetic modification has had success and research on the subject should be supported and intensified. For example, research using genetic modification has been responsible for the manufacture of synthetic human insulin through the use of GM bacteria approved by USFDA in 1982. Manipulation/modification of genes contributed to the production of a new type of experimental mice such as the oncomouse (cancer mouse) invaluable in medical research. Also, a genetically engineered vaccine for humans against hepatitis B was approved by the USFDA in 1987.

The biological techniques to increase global food supplies discussed in the previous sections are approaches to feeding populations that suffer from lack of food and from foods with low nutritional value. Additional tactics to increase food supplies locally, nationally, regionally, and worldwide include farming more arable acres/hectares, using sustainable farming practices, and protecting harvested crops by improving storage facilities (no access for vermin, dry). These will be considered in the following paragraphs.

Open Arable Land With Potential To Produce For Sustainable Agricultural Development

The Food and Agricultural Organization of the United Nations (FAO) defines arable land as land under temporary agricultural crops, temporary meadows for mowing or pasture, land under market or kitchen gardens, and land fallow for more than 5 years. Crops are defined in two categories, temporary and permanent crops. Temporary crops are both sown and harvested during the same agricultural year, sometimes more than once. There is less than a one-year growing cycle and they must be newly sown or planted for further production (e.g., maize/corn, rice, wheat, soy). Permanent crops are sown or planted once and then occupy the land for some years and need not be replanted after each annual harvest. According to FAO, they are mainly trees (e.g., oranges, apples) but also include bushes and shrubs (e.g., blueberries, blackberries), palms (e.g., dates), vines (e.g., grapes), herbicious stems (e.g., bananas), and stemless plants (e.g., pineapples).

Of the more than 13 million hectares of the Earth's land area, almost 5 billion hectares are arable terrain. Of the arable terrain, 1.5 billion hectares (30%) are in production and 3.5 billion hectares are under permanent pasture with 2.7 billion hectares of this having crop production potential. Areas containing grasslands, forests, protected areas, settlements (buildings and infrastructure), and industrial parks, for example, occupy potentially productive land and take it out of the agricultural development planning. Fourteen percent of the hectares in production have healthy soils. The other 86% suffer some degree of fertility loss from processes such as erosion, nutrient depletion, salination, and heavy metals pollution.[29] This results in lower crop yields and less quality crops with lower nutritional value. Annually, 5-6 million hectares add to the soil-degraded category.[30] Degraded soils can be reclaimed in many cases such as by replenishing nutrients (with agricultural chemical or with natural additives) and by washing out salts accumulated at root systems.

Globally, there are great differences in the amount of arable land in production, The highest percentage use of productive agricultural land is in Europe with 62%, followed by North America with 47%, Asia with 34%, Africa with 21%, and Australia/New Zealand with 10%. Clearly, Africa and Asia have to increase the area of potentially productive land as countries in these regions strive to feed existing populations and what are projected to be much larger future populations.[29]

In Africa, for example, since the 1950s more than 325 billion hectares of fertile farmland have suffered some form of soil degradation that has hurt food production.[31] At this rate, Africa's loss of agricultural productivity could be halved by 2030 even as its population doubles leaving 40% of the African population with chronic malnutrition and starvation. This is frightening and to some degree representative of what is happening in other parts of the less developed and developing world. There is hope, however. In less developed and developing countries, there are 2.5 billion hectares that have the potential to be productive, yet only about less than 1/3 are under cultivation.[29] If only another 1/3 of the uncultivated hectares are prepared, come into production, and are worked following sustainable farming norms, food crop yields will increase. This means that much of the chronic under-nutrition suffered by large numbers of citizens in less developed and developing nations can be greatly reduced or perhaps essentially eliminated. The greatest potential for agricultural expansion is in Sub-Saharan Africa and in South and Central America that together have 70% of the arable land with potential for production.

If farm productivity can reach levels to feed today's malnourished populations, the world will have made a great societal advance. If crop harvests can improve so as to be able to feed a future global population half again as large as today, agricultural planners with a team of agronomists, soil scientists, bio-scientists, politicians, and economists will merit a group Nobel Prize in Planning, Productivity, and Prosperity. To reach for these goals, planners have to teach and support practices that both minimize the loss of soils and counter soil degradation. Only by doing this can agriculturalists hope to attain a level of food production that will suffice to feed today's populations and a 40-50% increased global population in about two generations.

Practices to Reduce Soil Loss and Sustain Soil Fertility

Soils may lose fertility and hence productivity if they are degraded. Degradation can be caused by erosion, the physical removal of soil from farmland by flowing water, wind, and gravity. Most erosion worldwide is the result of soil exposure by vegetation removal, overgrazing, and damaging agricultural practices. In Asia, vegetation removal is dominant followed closely by agricultural activities and overgrazing whereas in Africa overgrazing is dominant followed by agricultural activities. If these actions can be controlled, the soil erosion problem in the regions becomes less threatening to food production for both existing and future populations.

Soil degradation may also evolve from a loss of nutrients (chemicals that are necessary to sustain plant growth), from salts that encase plant roots and prevent uptake of nutrients (salination), and from pollution by potentially toxic metals and other toxins. More than 3/4 of soils worldwide are degraded to a lesser or greater degree. The result is that farmland yields of crop per acre/hectare and crop quality are not as high as they could be if the causes of degradation were minimized and soil fertility maintained. Soil degradation reduces the global food supply and certainly will put great stress on growing populations. This is especially true in less developed and developing nations, unless soil degradation of their farmlands can be moderated and its effects subject to remediation.

Limit Erosion

There are several methods farmers can use to minimize soil erosion. Principal among these is to change tilling practices (preparing a soil for planting, sowing seeds, and raising crops) and harvesting methods. The preparation of a soil for sowing by ploughing breaks up the soil cover and exposes the soil to erosion by flowing water and wind. Conservative soil preparation uses equipment that slices and seeds with a minimum of rupture of a topsoil. This limits soil loss by erosion. Farmers in less developed and developing countries do not have the capital to invest in the slice and seed machines. This preferred soil preparation method would be used on large scale in these countries only if international organizations or NGOs make the machines available through a type of lend-lease program. This is unlikely given the 2008-2010 global economic condition. Nonetheless, other erosion reduction tactics can be used. For example, farmers can reduce soil erosion by half on gentle slopes by ploughing at right angles to the slope. Also, working farmland soil in the spring when vegetation cover comes in rapidly reduces soil erosion versus ploughing during winter when ruptured soil is susceptible to erosion for a longer time. When cropping on hillsides, terrace farming is the norm to prevent important losses of soil. The terraces should be short-walled to retain water and prevent runoff down hill that would cause large-scale erosion.

If breaches appear in terrace walls, they must be immediately repaired.

Farmers can maintain soil cover and limit erosion by planting crops that come in at different times during the growing season. This multi-species cropping has an added benefit besides erosion control. For example, mono-cropping can result in wholesale crop loss and economic stress for a grower if the single crop fails because of disease. Multi-species planting should be the *modus operandi* in any farming program. Also planting trees and other vegetation around farm fields will slow soil erosion during runoff and minimize wind driven soil erosion.

Maintain Nutrient Levels

Soils are depleted of nutrients as crops grow seasonally and year to year. The depletion will degrade a soil and reduce its productivity. This can be averted if nutrient replenishment is practiced. Nutrient depletion can be monitored by regular chemical analyses that can determine which nutrients need replenishment. The replenishment can be with man-made chemical fertilizers or by using organic farming techniques such as ploughing crop residues and compost into soils.

Reduce Chemical Degradation

Chemical degradation of soils comes mainly from the introduction of pollutants such as heavy metals from the atmosphere or from runoff where metals are mined or processed. It can derive also from natural sources such as volcanic eruptions and runoff from hidden (unexploited) mineral deposits. Crops can take up heavy metals from a soil during growth and pose a threat to consumers if the metals in foods bio-accumulate in body organs over time. The problem can be eased by eliminating the sources of the pollutants, and by planting crops that will discriminate against and not take up the toxic metals, or by remediation of the soil before it is used for cropping. Pollutants can also be pesticides, herbicides and other man-made biocide chemicals that adhere to field and tree crops. These can cause health problems if the crops are not thoroughly washed or peeled before consumption. Man-made fertilizers (nutrients) and biocides (weed, insect, other pest killers) have to be used judiciously to limit any entry to juxtaposed ecosystems. Each crop in a soil and in an area with given climatic conditions requires the application of different amount of agricultural chemicals per area of cropland. This means that a farmer or agricultural agent can calculate the optimal amounts of fertilizers and pesticides/herbicides to use to give maximum yields and best quality product while protecting a crop. This has the economic benefit of minimizing expenditures for the chemicals for a farmer or agribusiness. It benefits the food supply needed by today's populations and will augment the supply that will be needed for future generations.

Large quantities of harvested foods designated for distribution and use are lost from the global inventory each year. Much of the loss is in staples such as corn, wheat and rice. It is often the result of poor storage that allows a harvested commodity to be eaten and contaminated by animals (rodents) or spoil from moisture and bacterial activity. These food supplies need not be lost if proper storage facilities are used. The transport of food from where it is plentiful to where it is needed requires infrastructure for receiving, storing, and distributing the food. This can only be done with political compliance by governments whose citizens are the recipients of humanitarian assistance. Often time governments put up barriers to the import and distribution of humanitarian food aid such as has happened in recent years in Burma, Zimbabwe, and Sudan.

Fish - A Major Food + Animal Protein Source For Humankind

Fish is an important food source for people worldwide. Data for the years 1985-1997 show that fish consumption provided more than 2.6 billion people in Southeast Asia with at least 20% of their average per capita animal protein intake. Globally, food fish account for 15.9% of the per capita animal protein intake. The annual rate increase of food fish ingestion for the developing world is 3.8% but is -0.1% without China. The developed world has a declining rate of fish consumption of -1.0%. China, with an annual increase in fish consumption of 10.4% and a per capita intake of 26.5 kg/year is a rising global consumer but is far behind the world leader Japan. Japan has a fish intake of 62.6 kg/capita/year but an annual growth rate in fish consumption of only 0.2%.[32]

Most of the fish humans eat, low value and high value finfish, and crustaceans and mollusks, are captured from oceans and seas with lesser but important contributions from estuaries, lakes and rivers. However, the aquaculture contribution to the human fish diet is growing and now provides up to 40% of global fish production. The growth in aquaculture will accelerate if overfishing of the oceans continues and there are decreased harvests from many fishery ecosystems. We discuss overfishing and its threat to the world's food/protein supply in the next chapter. Aquaculture has been making up for smaller ocean harvests and greater demand, often with genetically modified species but not of large food fish that are much sought after such as tuna. There is a problem here in that we do not know what effect genetically modified species that escape their marine aquaculture confinement will have on wild fish when they interbreed.

In a similar way that nutrient replenishment is necessary to maintain soil fertility and productivity, fisheries need time for fish stocks to replenish. This has not been the case in all ocean fisheries because of the failure of the fishing industries from some countries to adhere to international treaties that set catch quotas of certain species. This is driven in part by substantial investments made in fishing fleet ships and processing and distribution facilities that have to be justified. Some governments provide subsidies to promote overfishing and over capacity. These subsidies must be reduced to support a rebuilding of marine ecosystems and their fishing stocks. Adherence to catch quotas will preserve fish species as a food source for a growing world population particularly in less developed and developing countries. Harvesting to sustainable quotas into the future will translate to long-term profits that would justify global fishing industry investments.

At their annual meeting in November 2009, The International Commission for the Conservation of Atlantic Tunas (ICCAT), a 48 nations group, reduced the Atlantic bluefin tuna quota from the 2009 limit of 22,000 metric tons to 13,500 metric tons for 2010 to help the bluefin tuna stock to recover. They also committed to a science-based catch level for 2011-2013 and a reduction in fishing capacity. In addition, the ICCAT lengthened the existing purse seine fishing closure for 30 days during the critical spawning period. Purse seine fishing involves a fishing net hauled between two boats that encircle a school of fish. These are actions late in coming because the bluefin stocks are now less than 15% of what they were before the onslaught of modern, large-scale commercial fishing. Nonetheless, these approaches should boost the recovery of the bluefin tuna stock but certainly would be more effective and yield results sooner than projected if the quota was cut further. Such a further reduction would achieve a high probability that the stock would recover within a generation or less. This would allow sustainable catches of substantial tonnages of the bluefin tuna in the future. However, European fishing nations (e.g., France, Greece, Italy, and Spain) opposed a greater reduction in the allowable quota.

Overfishing has not only diminished stocks of high value and low value food fish species but in some cases has interrupted the marine food chain so that some predators have to search out new prey to sustain their being. Only much reduced hunting or non-hunting of these species and time will allow them to restock and benefit humankind with their food/animal protein input to peoples' diets. Enforcement of fishing quota limits is lacking and some reports in the 1990s have us rushing towards losing 75% of global fisheries by 2025. However, in a 2009 report titled "Rebuilding Global Fisheries", Professor Boris Worm and his colleagues posit that fish stock rebuilding can work for marine ecosystems where fishing is controlled by individually tailored managerial practices.[33]

These authors recount that marine ecosystems fish stocks are in varying stages of sustainability ranging from stable, to declining, to collapsing, to rebuilding. They found that in some ocean regions there has been a reduction in exploitation rates because of implementation and following of management practices designed to rebuild fish stocks. However, many marine ecosystems are flirting with collapse because of lack of harvesting control in many regions of the oceans and seas. This notwithstanding, marine ecosystems can be rebuilt. Recovery is documented for 1% of the oceans and seas. Recovery is in progress for other ecosystems in decline or nearing collapse because their rates of fish catch have recently lessened and are below the rate necessary to achieve optimal sustainable yield. Professor Worm and his colleagues estimate that rebuilding is necessary for

63% of the fish stocks assessed in recent studies. This can be done if exploitation rates are lowered and other practices are followed that allow rebuilding of fish stocks. However, rebuilding ecosystem communities with their fish stocks to the degree that they can be harvested for sustainable yields takes time and hence is not without pain. Reduction in fishing has short-term socio-economic implications such as the loss of employment in the actual fishing phase, the selling to wholesaler phase, the resale to markets phase, or the resale to canneries. Lost income translates into less purchasing power and a weakening of local economies and falling government tax revenues.

Management tools used to reduce fishing exploitation rates vary according to the marine ecosystem to be rebuilt. Together, reductions in allowable catch and in harvesting capacity by diminishing the total fishing effort can bring down the rate of fish catch. This allows fish stocks that are depleted or nearing collapse numbers to rebuild. Areas closed to ocean fishing have the same effect. Restrictions on the sizes and characteristics of gear used in harvesting the oceans and seas allows for selectivity on the species captured and favors a reversal of the drop in populations of targeted fish species. The sharing of a fixed portion of a total allowable catch among fishermen is a management tool that is fair to fishermen economically and works against over exploitation. This maintains a sustainable fish stock and allows rebuilding to take place. These methods can be applied effectively when communities affected by them have input during planning and decision-making. Federal laws that make overfishing a crime with sharply defined mandatory punishments and fines for fishermen and executives of companies that break specific laws are powerful incentives against flaunting the law. Unreported and illegal fisheries catches in international waters represented 18% of the total catch from 2000-2003.

This problem must be addressed by the international community and offenders punished as crimes against nature (our living environments) and society.

Finally, it must be noted that food from the sea includes mammals...whales. Whale hunting by relatively few countries has put some species in the endangered category. The International Whaling Commission controls whaling and determines the whale species and the number of animals that can be taken from the marine ecosystem. Japan tried to get around whaling restrictions during 2008 by setting quotas of 800 to 1000 whales to be captured for "scientific research". Whatever so-called "scientific research" the Japanese propose to do to justify large scale whale hunting can be carried with far less a catch. Although the whale meat from the Japanese harvest is sold in markets and restaurants, the Japanese claims that their hunts of 800 to 1000 are for research can be considered an untruth. In this case, Japan is a developed modern industrial society with medieval standards when it comes to what is considered a traditional culinary delicacy...whale meat. On the other end of the reality spectrum, whales and seals hunted for food and furs for clothing and other uses by the Innuit people and other northern clime inhabitants do not endanger either species because the indigenous people kill only the number of animals that they need in order to survive.

Capability to Feed Present and Future Populations

We return again to a familiar theme to complete this chapter. Many governments of less developed and developing countries in Africa, Asia, and the Middle East are not meeting the clean water and nutritional needs of large numbers of their 2009 populations. How then can we, as an international community, work with these governments and supply safe water and necessary foodstuffs to populations sometimes half again as large in these countries in 2035-2050?

We have discussed actions that can increase the amount of clean water available globally and that can augment the world's food supplies. The real question is how large a population can be sustained with a reasonably good quality of life? Will an increased production and availability of water and food be enough to sustain a global population that is projected to stabilize at 9.5 billion humans or more by 2050? If the answer is positive then we must ask if governments have the will and politico-economic backing to create or improve an infrastructure to generate or receive these life-sustaining commodities and deliver them to where they are lacking? Some researchers do not believe so. They estimate that the Earth can sustain closer to 7-8 billion souls with a reasonable quality of life. Our recent history and existing conditions argue against success. Nonetheless, the threat of world disorder in the competition for water and food may focus public actions and government responses towards seeking a bettering of distressful national and global conditions.

If governments and international alliances do not respond affirmatively to implementing necessary changes in policies with regards to the very staffs of life, water and food, this portends badly for future generations. Failure to move to immediate actions raises a global specter of ugly competition for commodities that sustain life, especially in severely stressed populations and governments in "have not" countries. This situation could spark violent wars within a country and across national boundaries and ultimately threaten the well being of all nations. The future will reveal the answer.

AFTERWORD

Ideally, there should be immediate positive responses of the global community to use new technologies, biosciences advances, economic and sociological incentives, and enforceable legislation to solve problems of societal water and nutritional needs where they are lacking. Realistically, responses will be slow for three main reasons: first, a lack of economic resources; second, the need to train a cadre of personnel prepared to carry out planned responses to improve water and food supplies successfully; and third, a rush to industrialization by key developing countries (China, India, Brazil) plus Russia and Mexico.

As programs to solve sustenance problems are put into practice, they will help ease shortfalls of food and water that leave billions of people in many less developed and developing nations today with chronic health conditions, in poverty, and in an untenable subsistence mode. With clean water and sufficient nutritious food plus socio-economic benefits such as access to education and healthcare, citizens can be in a position to make substantive contributions to their country's development.

Whether this will hold true for greatly increased numbers of people in the future will depend on the ability to ramp up sustainable practices to increase water and food supplies to nourish them. It depends as well on the international will to confront other factors that affect their environments now and will do so in the future, sometimes with heightening effects. These include naturally occurring and generally localized physical hazards (e.g., earthquakes, volcanoes) plus disasters that bear the imprimatur of human activities that are dangerous to populations and environments worldwide (e.g., global warming). We will discuss these issues in the next chapter and examine their complex relations in an expanding populations scenario, for less developed and developing countries.

CHAPTER 5. NATURAL AND ANTHROPOGENIC EVENTS THAT THREATEN GROWING POPULATIONS TODAY AND IN THE FUTURE: FAILINGS AND SUCCESSES IN COPING WITH THEM BEYOND SOCIAL, ECONOMIC, AND POLITICAL FIXES

Natural hazards and those disasters that evolve from human activity or lack of human response to them affect how we deal with contemporary environmental problems. They direct our planning for future issues that will be caused by expanding populations and higher population densities. As stated in previous chapters, demographers estimate that by 2050 the number of people on earth will be about 9.5 billion but could reach 11 billion, or about 40% to 60% more than in 2008. As populations grow and people densities increase, it is likely that natural and human-originated disasters will have greater negative consequences for future generations.

The onslaught of these events, whether limited, regional, or extensive in their reach (e.g., earthquakes, acid rain, pandemics, or global warming), or physical, chemical or biological in nature, will stress out citizens. The populace will put pressure on governments to take steps to prevent the same degree of injury, death, and destruction suffered in past events from similar future happenings. Proactive planning commissions can author regulations that mitigate the effects of such events if adopted and enforced. Action now, backed by political will and economic resources, and taking into account cultural norms and religious beliefs, is essential to secure the well being of humans on Earth today and in the future.

NATURAL DISASTERS

Several biological and physical primary events happen naturally without any human influence. They shock populations and can bring sickness, injury, and death to them and can damage and destroy their living and working spaces. The dangers to life and property of some of these disasters are compounded by secondary hazards they can trigger that multiply harmful impacts on populations. The effects of the primary happenings and secondary hazards they spark can often be alleviated substantially by means achievable within the ability and the volition of government managers who direct mitigation programs to carry them out.

Biological

There are biological threats to populations beyond the failings to nourish the human body with water and food as discussed in earlier chapters. These are menacing diseases. Some could put the health status of the earth's growing populations in great danger, especially if the source of a disease remains active and if a therapy is not available so that the disease becomes epidemic or pandemic. Epidemics spread from contagious diseases such as typhoid, cholera, or tuberculosis. They tend to affect many individuals in a population, community, or region at the same time. Pandemics occur over a wide geographic area, may infect an exceptionally high proportion of the population, and are transmitted by direct contact between life forms (e.g., malaria, dengue fever, HIV/AIDS, influenza).[34]

With proper hygiene and sanitation, epidemics of diseases such as typhoid and cholera can be avoided. However, if typhoid or cholera is contracted, people can be treated. The spread of the disease can be slowed and halted once the source is identified and isolated. During any given year relatively few die from afflictions such as these because treatment is generally available. Tuberculosis, however, can be deadly. During 2007, two million people died of tuberculosis and 9 million people carried the disease. Only a two-year treatment on medications by mouth (up to 10 pills daily) and injections can cure multi-drug-resistant tuberculosis, the worst form of the disease.

In an earlier chapter we discussed the "black death" pandemic that killed 28% of the Chinese population from 1318-1338, 1/3 of Europe's population during the period 1347-1352, and reduced the world overall population by 17%. We also cited the Spanish flu pandemic that killed an estimated 50 million people during 1918-1919. These were major population reducers. The possibility of suffering grand killer pandemics in today's societies is mitigated by several factors. These start with rapid and open communication among global health services about the possibility of the onset of a pandemic. Further, there are continuing improvements in healthcare technologies and in medication therapies. To this we can add advances in disease control methods and medicines/vaccines plus the capacity to quickly move medicines and vaccines from one part of the world to another where they are needed. This reduces the potential of a disease to spread and to be fatal. Thus, population losses to diseases in contemporary societies are not likely to be as severe as in the past. For example, malaria is pandemic and affects more than 515 million people annually killing 1-3 million people. Some 90% of malaria related deaths are in sub-Saharan Africa and the majority of these are young children.[35] Positive advances to prevent malaria include the use of bed nets treated with the mosquito repellent permethrin, a pyrethroid insecticide incorporated into a polyethylene fiber netting that can last up to five years through 20 gentle washings. This complements mosquito control programs for breeding sites. Singapore has a successful program with strictly enforced laws that prohibit standing water (breeding site) at construction locations and at home properties. Where treatment is made available to people afflicted with malaria, the number of deaths from the disease is greatly reduced.

Dengue fever infection affects at least 50 million people annually in Africa and in the tropics (e.g., Australia, Southeast Asia, Central America, the Philippines, the Indian subcontinent, the Caribbean, and South America). Of the 50 million persons who contract dengue fever, 2.5 million are at risk of death. The same techniques that are used in the fight against malaria can be applied to prevent dengue fever: mosquito control and use of bed nets treated with mosquito repellent. Research is advancing towards the formulation of a vaccine against the disease.

The HIV/AIDS pandemic is spreading although there are methods to prevent its advance and medical therapy to treat and maintain a good quality of life for those infected with the HIV virus. The therapy is limited to those who can afford it or who are fortunate to receive the therapy through their healthcare services. It is sad that the Health Minister of South Africa backed by President Mbeki did not believe that the HIV virus was associated with AIDS. He advocated that eating garlic and beetroots would fend off HIV and was negative on the use of antiviral mediations recommended by the World Health Organization and medical professional organizations worldwide. When the Deputy Health Minister contradicted her boss in at international conference in 2007, President Mbeki fired her. Sadly, the leader of Gambia followed the garlic-beetroots path. Thirty-three million people are infected with the HIV/AIDS virus worldwide with 22.5 million cases in Africa led by 5.3 million in South Africa where almost 1000 people die and 1400 contract HIV daily. Globally, 5700 die from AIDS and 6800 are infected with HIV daily. During 2007, 2 million people died of AIDS but the pandemic worsened as 2.7 million contracted the HIV/AIDS virus.[36] Certainly, education of populations at risk on how to limit the chance of contracting the disease by the use of condoms is important but is clearly not reaching many of the susceptible populations. It is especially important to reach women of childbearing age who could become pregnant and pass the HIV virus to a fetus. It was disheartening during Pope Benedict's visit to Africa during March 2009, to hear him advise Africans not to use condoms. This can cause grave problems for the African populace in many nations in two ways if believers follow the advice. First, it will lead to a more widespread and faster spread of the HIV virus between adults, teen populations, and to fetuses carried by women who have contracted the virus because a condom was not used. Second, if Pope Benedict's doctrinaire call for the non-use of condoms is followed, it will contribute to the burgeoning population growth in Africa that today is the site of poverty, sickness, under-nutrition, starvation, and conflict. It will also result in greater numbers of AIDS victims. If a majority of the 815 million Africans do not have a reasonable quality of life in 2009, how can a doubled population by 2040 of 1.6+ billion souls be expected to survive with a lesser quality of life and one racked with disease? The 2040 African population will exceed the 2040 populations of China and of India.

The epidemic and pandemic sicknesses that afflicted our populations during 2009 will likely affect more of the global populations as they grow by close to half or even more during the next few generations. As emphasized repeatedly, the people who will suffer most from debilitating or killer diseases will be in less developed and developing countries. The deaths from disease globally will not cause great changes in numbers of the Earth's citizens. Deaths from diseases, deaths from wars, and natural mortality rates, will be more than made up for by the annual global population increase.

Physical

There are several physical hazards that threaten existing populations. They present a greater threat to growing populations because of higher people densities in areas susceptible to a natural disaster. These are earthquakes, volcanoes, floods, landslides, and extreme weather events such as high-energy tropical storms (hurricanes, typhoons, monsoons), and drought, plus triggered secondary hazards that they generate. There is no halting the occurrence of natural physical hazards. All can maim and kill people, and destroy property and infrastructure. However, the number of persons who are injured or who die is small relative to national or global populations. Nonetheless, they are overbearing for the families and communities that are affected.

Earthquakes

Earthquakes are sudden events that generally last from 10s of seconds to a minute or more and mainly affect localized or regional areas. The sudden shaking, jarring, rolling motions they generate at the earth's surface can bring down buildings and disrupt infrastructure when the intensity of motions is strong enough. Aftershocks add to the horror of an initial violent earthquake motion by collapsing buildings and infrastructure weakened by the initial shock. Earthquakes do not kill people.

Collapsing buildings kill people, as do fires from ruptured gas lines and pipes carrying water needed to fight the fires. People also die from an inability of disaster response teams to arrive at distressed areas rapidly or from a lack of equipment to search out and extract them from collapsed buildings. People suffer as well from delays in providing food, water, medicines and shelter where desperately needed, and in creating sanitation sites (latrines) and supplies. When thousands die as the result of an earthquake, it is essential to bury bodies as quickly as possible to avoid their contact with water so as to prevent water-borne disease (gastroenteritis) that can originate from them.

The effects of earthquakes are expected to be most severe at locations with large populations in high densities...urban centers. This was evidenced in Mexico City during a 1985 earthquake that is purported to have killed 23,000 citizens, mainly from collapsing buildings and infrastructure elements. Similarly, the earthquake that brought injury, death, and destruction to the Kobe-Osaka urban metropolitan centers in Japan during 1995 killed about 6500 people, many from the collapse of heavy tiled ceilings and from fires that resulted by ignition from broken lamps and toppled stoves. The fires could not be contained because of ruptured water pipes. Fires ravaged closely spaced wooden homes and buildings in the heavily populated neighborhoods. This was the worst loss of life from an earthquake in Japan since the 1923 Kanto disaster that killed well over 100,000 people in densely populated Tokyo, Yokohama and surrounding areas. In 2008, the Sichuan earthquake in China released about 30 times more energy than did the Kobe-Osaka earthquake. Almost 70,000 died, 375,000 were injured, 20,000 were missing, and 5-11 million people were made homeless. Densely populated centers and shoddy construction were responsible for many deaths. The largest loss of life from an earthquake in China was in 1976 when the Tang-shan event killed almost 250,000 people. It is not only heavily populated areas alone that suffer great losses from death, injury, and mass displacement of populations as evidenced by the 2007 earthquake in portions of Kashmir province governed by Pakistan and India, respectively. Here, about 80,000 people died from collapsing homes and deadly landslides in towns and villages spread over large areas of this difficultly accessible Himalayan region. Millions were displaced from their homes. The largest number of deaths that resulted from an earthquake in recent years was from a tsunami generated by a high intensity earthquake on the ocean floor offshore of Sumatra, Indonesia at the end of 2004. The tsunami, and flooding and landslides it caused, killed at least 230,000 people in four countries bordering the Indian Ocean. Most of the dead lived in highly populated Indonesian coastal areas and in coastal towns in Thailand, Sri Lanka, and India.

As yet, scientists cannot predict exactly where and when an earthquake will occur. Nor can they predict the magnitude or the motion intensity one will have. However, as scientists continue to review data from past earthquakes and evaluate physical, chemical and biological observations and measurements from them, concepts for earthquake prediction are being refined. Statistical analyses now allow generalized predictions for locations where earthquakes have a high probability of occurring, a probable range of magnitude that can be expected, and an estimate of when, in a range of decades, one is likely to occur.

Unfortunately, this does not protect threatened populations from the effects of earthquakes that do happen. A reduction of the maiming and killing of populations now and in the future is

Food was being distributed (Haiti imports 60% of food needs), bottled water was being distributed, and water purification units were brought in a put into service at locations sited by Haitian government representatives. Much of this was possible through the use of helicopters from US Naval vessels and from the Dominican Republic. Family members and then Haitian government people were dealing with the heart-rending task of retrieval and burial of the dead. Care must be taken here because human contact with bodies that have borne infectious diseases (e.g., hepatitis C, tuberculosis) can pass them to those handling the bodies. All in all, there has been a lack of preparedness for responding to large magnitude natural disasters by weak governments and their institutions past and present. As this is being written there is no rain but there is a fear of landslides that can be triggered by aftershocks, or by rainwater seeping into the deforested hillsides. Landslides here can destroy structures that remain intact, can block roads to hinder relief efforts, and can dam rivers making flooding another hazard to cope with.

Preparedness to cope with earthquakes or any natural disaster is a key to saving lives, minimizing injuries, and turning chaos into order. On February 27, 2010, almost six weeks after the Haitian earthquake, an 8.8 magnitude earthquake rocked Chile's Central Valley at 3:34 in the morning over a latitudinal distance of about 375 miles. This earthquake released 300-500 times more energy than the Haitian event. There were dozens of aftershocks by March 1 with magnitudes ranging from 4.9-6.9. By March 5, the government reported that the earthquake affected 2 million people, that about 500 people died, and that at least 500,000 homes were badly damaged or destroyed. In contrast to the situation following the Haitian earthquake, the Chilean government was immediately functioning and prepared to respond to the happening. Search and rescue teams were in the field quickly. The dead were to be identified and returned to their families for burial. The government used the army and air force to bring in food, water, blankets, tents, and medicines by air or overland and to set up distribution points for these essentials. Local police and army troops established order and minimize looting of markets for food and water. Criminals looting stores were apprehended. Field hospitals were set up. Heavy equipment was available to remove rubble where necessary. Communication was re-established in many areas. The President and her assessors flew over the devastated regions in a helicopter to view the extent of damage and determine immediate needs to help citizens.

The number of dead from the Chilean earthquake will increase but will be a fraction of one percent of those in the Haitian earthquake. This is in grand part because Chile has a building code that was modified after a devastating earthquake in 1960 and that mandates earthquake resistant construction. There is no earthquake proof construction. The code has been constantly modified as new norms for construction are generated by engineering data garnered from the analyses of structures and infrastructure response to earthquakes worldwide as they occur and are analyzed. As a result there was no wholesale collapse of building as there was in Haiti especially for structures built after 1960. This saved many, many lives. Certainly, the greater depth of the Chilean earthquake focus (21.7 miles) and greater distance of the epicenter from population centers (about 100 miles) with respect to the Haitian earthquake focus and epicenter and proximity to Port-au-Prince, were important in keeping the number of deaths and amount of destruction down. Airports were functioning to receive aid requested from donor countries such as temporary bridges, field hospitals, satellite phones, electric generators, water purifications systems, field kitchens and dialysis centers.

One instance when the preparedness failed was the mistake by the Navy to immediately activate the Chilean tsunami warning system after the quake hit. More than half the dead were killed by tsunamis. Nonetheless, many lives were saved from the killer tsunamis that reached heights of 10 meters by port captains in several coastal towns who did activate the system. This gave people 23-30 minutes to follow evacuation routes inland. Seafronts and entire towns were demolished by the tsunamis but the warnings saved 100s of lives.

Volcanoes

Volcanic eruptions are also sudden events, but events that can last for varying periods of time, from days to weeks and often as longer sporadic eruptions. A non-violent eruption may emit lava that will flow down slope following the topography and destroying whatever it rolls over. Lava flows generally move slowly on slightly sloping terrain and people in the path of a flow walking at a rate of 3 mph should be able to move out of its path to safe higher ground. Violent eruptions emit ash that can rise ten miles or more into the atmosphere, glowing rock bombs, and fiery ash flows (nueés ardentes). They can trigger secondary events that present a great threat to life and property such as fires, lahars (hot mud flows), and landslides. Violent eruptions have killed tens of thousands in the past when scientists did not yet have prediction tools that could signal the real possibility of an imminent eruption. People also died because they did not heed evacuation warnings that were given. In addition, grand undersea volcanic eruptions in the past have generated killer tsunamis although these have been rare happenings.

Fortunately, scientists have made remarkable advances in volcanic eruption prediction using observations and measurements based on the premise that magma (molten rock) is moving in the subsurface and rising towards the surface to cause an eruption. These can be detected and include the following: 1) emissions of wisps of smoke/gases and increases in their volumes and frequency; b) a bulging flank of a volcano; c) changes in chemical composition of gases being emitted; d) increases in temperature of soils and lakes around a volcano; and e) changes in the signal of seismic activity before an eruption. Topographic analysis and meteorological data reveal likely paths that ejected materials from explosive eruptions will take. Scientists have established risk zones around volcanoes using these data and those from past non-violent and violent eruptions. On the basis of precursor signs to eruptions and information cited above, civil defense officials can put together warning systems and when they deem it necessary, can call for a mandatory evacuation from areas at risk close to or several miles from a volcano. This will avoid injuries and save lives although property may be lost. Many millions of people live on sides or near the bases of volcanoes or close to them to farm rich volcanic soils. Many more will inhabit the at risk volcanic areas in the near future as populations increase. Thus, the warning systems and adherence of the populace to them is essential to peoples' security against injury or death from volcanic eruptions and their triggered hazards.

Floods and Landslides

Floods originate naturally from too much rain in short periods of time (flash flooding), continuous rain over an extended period of time, early and rapid snowmelt during the late winter or early spring season, or from the failure of ice jam dams. Floods can be the result of human activities as well and this aspect will be discussed in a later section. When the natural and anthropogenic causes of flooding occur together, the invasion of living and working spaces can be disastrous. Fortunately, flooding can be predicted. Scientists use equipment to obtain real time data and calculate what is happening in various sectors of streams and rivers in a drainage basin in terms of the volume of water input and the velocity at which the water is moving downstream. With these data they can accurately compute when flooding will take place in a given area and how high the floodwaters will reach above bank level onto a flood plain and beyond. This allows time to warn citizens at risk so that they can move home contents to a second floor (if they have one), pack up their valuables, family treasures, and important papers, and leave the flood zone. Of course, the magnitude of some floods will destroy homes and other structures in and out of a flood plain but property can be replaced, lives cannot be replaced.

Flood control techniques are used in many areas to safeguard citizens, especially in heavily populated areas. They also serve to preserve farmlands and their crops. These include location and construction of dams, levees, and floodwalls, and when financially possible, a deepening, widening, and/or straightening of stream or river channels. This allows more water to flow through an area more rapidly. Obviously, flood control structures have to be maintained and upgraded as necessary to assure their integrity and water containment capabilities. In extreme cases, where a town has been ravaged repeatedly by flooding, the town can be moved to a nearby site well above the floodplain. Construction on floodplains can be minimized using political and economic techniques. Zoning laws can prevent construction. Banks that will not grant loans to build on floodplains, and insurers (including governments) that will not write policies against flood damage will also limit floodplain development for other than agricultural projects.

Landslides, also called landslips, take place naturally when the force of gravity pulling down on a mass of soil and/or sedimentary rock on a hillside overcomes the frictional force keeping a slope in place. Continuous rain that seeps into a susceptible slope increases the weight of soil/sedimentary rock and thus the pull of gravity on the slope. The seeping water invades friction planes and decreases the gravitational force holding the slope in place. When the weight of the mass being pulled down by the force of gravity overcomes the frictional force keeping a slope in place, there is a landslide. Although common on steeper slopes, landslides can happen on slopes as little as 15 degrees. There is no predicting when or where a slope will fail but in an area with a history of such failures, sustained rains can be a precursor to setting up landslide conditions. Geologists inspecting a region with varying topography can determine where landslides have occurred in the past and where they are likely to occur in the future. With such information, local governments can warn citizens against building in a landslide prone area unless they include stabilization control methods in their planning for a construction site, be it for a home, business, or infrastructure.

Stabilization methods that have been used are to seal a slope to prevent seepage of water into it as has been done in Japan, or to put in a drainage system that collects and carries seepage away from the slope is common practice globally. Retaining walls with openings that allow water to drain out of a slope are used in residential neighborhoods. Certainly, undercutting the base of a slope for a road or structure, and/or overloading the head of a slope by putting in a road or large structure can result in an increased gravitational pull downward. If the pull of gravity exceeds the friction force resisting a movement, the result will be a slope failure. Other human-created conditions can bring about landslides.

Growing populations in the future are likely to inhabit areas susceptible to landslides unless planners can use political and economic disincentives to prevent them from doing so. This means passing legislation with the mandate that construction sites must have landslide reduction and resistance methods built into their projects.

High Energy Tropical Storms

About 60% of the global population lives within 200 km (124 miles) of a shoreline, in urban centers, towns, and villages. Many of these locations are at the mercy of high-energy tropical storms known as hurricanes in North America, typhoons in Asia, and monsoons in the Indian subcontinent. The number of people moving to these areas has been high, often for reasons related to economic, healthcare, and education conditions their population centers offer. The number will increase in the future because of population expansion, especially in less developed and developing nations. Some demographers project that 2.75 billion people or more than 1/3 of the projected 7.5-8.3 billion people world population in 2025 will inhabit a 100 km (62 mile) deep coastal zone.[37] It is likely that more than 3 billion people will inhabit this zone in 2050 when the world population is estimated to reach 9.5 billion people.

The heavy sustained rains from tropical storms can cause flooding inland. High winds associated with the most violent of them can drive storm surges inland like a mini-tsunami, down trees and power lines, and rip roofs off of homes and other structures or collapse them. In special cases as with Hurricane

Katrina in New Orleans, hurricane winds can drive inland waters against levees, build up pressure against them so that levees are breached and water behind them can flood areas the levees were designed to protect. To protect its principal economic center from typhoon driven storm surges, China has built tall seawalls (6.63 m high [22 feet]) along the coast of Shanghai to protect the metropolis. The technological and engineering leader in terms of protection from high-energy storms is the Netherlands. The Netherlands has constructed tall gates that can be put into place by electrical power, and walls (dikes, levees) to protect the topographically low country from storm surges. To cope with this and rising sea level, the Dutch are evaluating floating architecture or structures that are built on foundations that move up or down in response to rising or falling water levels for farms, commercial parks, and towns. They have built small communities in a pilot study to evaluate the practicality and economics of continuing with such projects. This approach is coupled with the creation of large area temporary storage pools to retain excess water from rain or wind driven storm surges.

Meteorologists can track tropical storms, determine their wind speeds, estimate dangers to populations, and give warnings with sufficient time so that people can evacuate if needed. Property is destroyed, but lives are saved. Whether evacuations can be carried out successfully depends on the preparedness of governments in the coastal zones affected and the rapidity with which they act. As more population relocates into coastal urban centers, greater numbers of people are at risk from events associated with high-energy tropical storms. Shanghai is an example of considerable growth mainly due to demographic migration from other urban centers and from distant rural areas in order to find better economic opportunities. The population of Shanghai city proper was 12.2 million in 2002, grew to 13.6 million by 2006, and reached 15.5 million in 2008. In addition, there was a floating population in 2008 of more than 3 million people. If a storm surge were to overtop and/or breach the Shanghai seawall, many lives could be lost. Planning and preparedness are the keys to successful protection of high-density populations in coastal habitats.

Drought

A natural physical event that will likely have a strong negative impact on the growing global population is drought, a shortfall of precipitation during an extended period of time. Droughts are often seasonal. As such, populations can plan for them by storing a three-month or longer supply of water in surface reservoirs or underground sites. However, if a drought lasts far longer than planned for, populations and their economic ventures can be in dire straits. In recent years droughts have had longer durations. Many countries and continents in 2008/2009 suffered extreme drought conditions that are historically their most severe droughts. Major food producers such as Australia, China, the United States, and Argentina fall into the historically most severe drought category. Crop productivity declines without sufficient water. For example, South and Southeast Australia have suffered an extended drought over 7 years. Rice production fell 98%. The lack of rice for export from Australia during 2008 put pressure on importing nations to seek rice elsewhere and drove the price up two and three times. This created hardships for populations of millions for who rice is the principal staple. It brought on violent protests in several importing countries (e.g., Indonesia, the Philippines, Thailand, Egypt, Ethiopia, Yemen, Ivory Coast, Uzbekistan, and even Italy). The drought damaged the Australian agricultural economy that depends on rice, wheat, wine, sheep products, and other agricultural exports. The situation in terms of rice production can worsen as farmers switch to less water-intensive crops such as wheat or wine grapes. Wine grapes production has increased at the expense of rice because it is more profitable. An acre of wine grapes can yield a pretax profit of $2000 but the pretax profit from an acre of rice is only $240. This portends badly for maintaining rice production in Australia even though a kilo of rice will sustain a person's food needs whereas a liter of wine (a kilo) will not. Changes in agricultural production from staples to high value crops are bad portents for growing global populations.

The stocks of foodstuffs of the major trading countries (Australia, Canada, the United States and the European Union) are on a precipitous decline. They have fallen from 47.4 million tons in 2002-2005 to 37.6 million tons in 2007 to 27.4 million tons in 2008.[38] This reflects the food situation in many countries and continents. As agricultural production of staples and other agricultural products falls, and populations in less developed and developing nations continue to grow, economic reality dictates that there will be higher prices for foodstuffs. For this reason, and the problems of food distribution from donor nations and NGO groups, 100s of millions more people will suffer in addition to the more than a billion people in 2009 already suffering from undernourishment and starvation.

Heat Waves

Heat waves are periods of unusually high temperatures often with high humidity that cause uncomfortable weather. These generally last 3-5 days to weeks. Heat waves can cause death and are most dangerous when they occur in densely populated cities. They abet urban conditions for the formation of killer smog. Cities are heat islands where daytime high temperatures may abate little during the night because concrete and asphalt absorb heat during the day and release heat at night. Heat waves can be deadly because of the reaction of a body to an extended period of high temperatures and because of the settings they create for wildfires. The aged, the very young (<4 years old), and people with underlying health issues are most susceptible to death by heat waves.

During July 1995, Chicago (USA) suffered through a 5-day heat wave with temperatures that reached a high of 109°F (43°C) and did not fall below 90°F (32+°C). There were 739 heat-induced fatalities. The heat problem was exacerbated by a power failure that left 49,000 homes without electricity to power air conditioners. The city was not prepared to cope with the heat wave. The government did not easily identify locations where people were at risk, had problems supplying fans and drink to people at locations when they were identified, and did not have a plan in place to efficiently move people to cooling centers as needed. City officials learned from this event. When another heat wave engulfed Chicago in 1999, the city was prepared with its Extreme Weather Operations Plan and the heat-induced death toll was down to 103 people. Heat waves will be discussed further in a following section on Effects of Global Warming.

THE HUMAN HAND MENACING POPULATIONS NOW AND IN THE FUTURE

Humans have a principle role in creating disasters, causing disasters, allowing disaster conditions to develop, and enhancing the effects of disasters. Some of these extend to global populations, others affect regional or national populations, and still others have local effects. This is obvious when we examine historical and contemporary happenings that brought pain and suffering to populations and the diminished vitality of ecosystems that nourish life forms on our planet Earth. This review will emphasize threats that face future generations as populations continue to expand at a rate that some believe cannot be sustained nutritionally, sociologically, economically, or by the finite resources of the Earth.

Wars/Conflicts

Beyond the actual number of deaths and percent of the global populations they represented, only world wars have approached or exceeded deaths from the "black" plague (1318-1352), mainly in Asia and Europe, and from the Spanish flu (1918-1919). World War I casualties totaled more than 38 million people with 16.5 million dead and 22 million wounded. Of the dead, 9.7 million were military and 6.8 million were non-combatants. World War II casualties are estimated at 72 million people of whom 25 million were military including 4 million prisoners of war and 47 million were civilians including 20 million who died. Most of these 20 million victims were gassed to death and then incinerated in ovens, shot to death in large groups and buried in mass graves, killed during bombings, or died from war-related diseases and famine.

War is defined as open and declared armed hostile conflict between states or nations as they struggle or compete for a particular end. However, in recent decades, terrorists groups with overt and covert state support have carried out wanton acts that maim or kill children, women, and the aged, as well as military foes in an effort to reach a sponsoring nation's political goal without the sponsor risking a nation to nation war. The terrorists then are surrogate players in international political battles. Wars in the ancient and historical past have been fought and are being fought today for the same basic reasons. These are ethnic hate, religious zealotry, natural resources (e.g., high value metals, diamonds), a desire to acquire land, a desire to rob a country of its riches, a desire to conquer other peoples to use their intellect or to enslave them, and a desire to reverse the results of such wars fought in the past for the fore mentioned reasons. As noted in the previous paragraph, many military personnel and civilians die and many others are wounded. As war machines have modernized, more non-combatants have died and are dying than are combatants. Sometimes this is the result of combatants fighting from densely populated neighborhoods and not allowing non-combatants to leave the danger zones under the assumption that their foes will not return fire there. It may be from using non-combatants as human shields. It may be from targeted combatant leaders living among the populace so that when the target is hit with artillery shells or guided missiles, innocents also die. Other times, wanton killing of civilians is to exact revenge for earlier atrocities in a war, or from pure ethnic or religious hatred, or from a psychopathic mind set of the urge to satisfy a need to maim or kill anyone at anytime.

As populations grow, competition for living space and natural resources will grow and this could lead to war, especially where the essences of life and a nation's existence and quality of life are at stake. Clearly, water is an essential natural resource as are fertile land to grow food and oil as a major energy source. Populations will grow for at least two generations and population density will increase with growth as well as from with demographic moves. The moves are to locations where there is access to the basic necessities of life: water and food, and where there is availability of healthcare, education, and opportunities for employment. As one result, the probability of high civilian casualties will be great in the event of war. Indeed with bombings instigated by terrorists, civilian casualties matter not as the aim is to try to disrupt a government's rule of law and cause population displacement. There has been talk of terrorists planning attacks using dirty nuclear bombs, toxic chemical gases such as sarin, and biological weapons such as air-borne anthrax or poisons in water supplies. Such attacks are directed not to incapacitate or kill a military force but rather to harm or murder civilians.

Overfishing

The capture of food fish from the oceans at a rate faster than the fish can spawn and reach maturity hurts existing coastal populations in less develop and developing nations that depend on fish as a staple and important animal protein source in their diets. The hurt will worsen as future generations add large populations that could otherwise be nourished in part by food fish. At present 28% of the ocean fish stock has been lost to overfishing. Some estimates signal that the oceans will lose over 70% of the existing species by 2048 as a result of overfishing and the loss of biodiversity (food chain disruption), and from pollution unless governments worldwide counter these actions.[18]

In a similar way that nutrient replenishment is necessary to maintain soil fertility and productivity, fisheries need time for depleted fish stocks to replenish. This has not been the case in many ocean fisheries because of the failure of fishing industries in some countries to honor international treaties that set catch quotas for certain species. This is driven in part by substantial investments made in fishing fleet ships, processing plants, and distribution facilities that have to be justified. Some governments provide subsidies that promote overfishing and over capacity. These subsidies must be reduced so as to curtail wholesale exploitation of marine ecosystems and thus give fish and other life forms in the ocean food web time to rebuild their communities. Adherence of fishermen to catch quotas will preserve fish species as a food source for a growing world population particularly in less developed and developing countries that border oceans, seas, and major lakes. Harvesting to sustainable quotas into the future will translate to long-term profits for the global fishing industry that would justify its investments.

Overfishing has diminished stocks of high value and low value food fish species and in some cases has interrupted the marine food chain so that some predators have to search out new prey to sustain their being. Only non-hunting of selected species and time can allow them to restock and benefit human-kind with their food/animal protein contribution to peoples' diets. Enforcement of fishing quota limits is lacking. Loss of a large number of ocean fisheries to help feed a possible stabi-lized population of 9.5 billion persons is troubling for inter-national organization planners concerned with food supplies. However, in a previously cited 2009 report titled "Rebuild-ing Global Fisheries", Professor Boris Worm and his colleagues posit that fish stock rebuilding can work for marine ecosystems where fishing is controlled by individually tailored managerial practices.[33]

These authors recount that marine ecosystem fish stocks are in varying stages of sustainability ranging from stable, to de-clining, to collapsing, to rebuilding. They found that in some ocean regions there has been a reduction in exploitation rates because of implementation of management practices designed to rebuild fish stocks. However, many marine ecosystems have not experienced a slowdown of overfishing and are flirt-ing with collapse over extensive regions of the oceans and seas.

Nonetheless, marine ecosystems can be restored. Recovery is documented for 1% of them. Rehabilitation is in progress for other marine ecosystems in decline or nearing collapse because the rates of fish catch from them has lessened in recent years as the result of adherence to quotas so that the fisheries can ultimately rebuild enough and allow optimal sustainable yield. Worm and his colleagues estimate that rebuilding is necessary for 63% of the fish stocks assessed in recent studies. This can be done using management practices discussed in the following paragraph. Re-establishing vital ecosystem communities with their fish stock to the degree that they can be fished for sustainable yields takes time and hence is not without socioeconomic pain. Reduction of fishing has what the industry hopes to be short-term implications such as the loss of employment for fishermen, wholesalers that depend on them, market fishmongers, and cannery workers. Lost income translates into less purchasing power, weakened local economies, and a loss of government tax revenues.

Management tools used to curtail overfishing vary according to the marine ecosystem in distress. Together, reductions in the allowable catch and in the harvesting capacity diminish the total fishing effort and will lower the rate of fish catch in marine ecosystems. This relieves pressure on fish stocks that are depleted or nearing collapse numbers so that they can rebuild. The same effect is achieved when national laws or international treaties designate ocean areas as closed to fishing. Restrictions on the sizes and characteristics of fish capture gear allow for selectivity of the species taken and favor a reversal of the drop in populations of depleted fish species. The sharing of a fixed portion of the total allowable catch among fisherman is a management tool that is fair to fishermen economically and is a deterrent to overfishing. Used judiciously, these tools keep sustainable fish stocks sustainable and also allow time for depleted fish stocks to reestablish themselves. These methods can be applied most effectively when communities affected by fishery controls have input during planning and decision-making. Federal laws that make overfishing a crime with clearly defined mandatory punishments, including prison time and fines for fishermen and company officials that break them are powerful incentives against flaunting the law. Unreported and illegal fisheries catches in international waters, the epitome of overfishing, represented 18% of the total catch from 2000-2003. This presents a grave global problem in terms of sustaining fish stocks. This issue must be addressed by the international community and perpetrators subjected to punishment for crimes against nature (our living environments) and society.

Floods and Landslides

Flooding and landslides are natural local or regional occurrences. They can also be secondary (natural) events triggered by earthquakes and volcanoes. The damage that these natural hazards inflict on people and property can be intensified by human decisions and actions. These can include: 1) allowing construction of homes, work places, and sensitive infrastructure on a 100-year floodplain (the water reach above bank level expected to occur once every 100 years); 2) no flood control methods in place; and 3) building permits that allow the overloading of a slope with the weight of homes, roads and other infrastructure or undercutting a slope for road building, actions that reduce the resistance of a slope to landslides. These classes of disaster are less dangerous to life than those discussed previously. However, their effects could be more dangerous for future generations because of population growth in countries where citizens do populate areas subject to floods and landslides but lack flood control or warning systems.

Subsidence

Another type of mass movement that can cause problems for populations is subsidence of an area of the earth's surface. Subsidence takes place in areas of sedimentary rocks such as sandstones that are porous and permeable. Porous and permeable rocks can hold fluids in interstices (porosity) and transmit the fluids (permeability). This is unlike the other major classes of rock called igneous rock (e.g., granite) and metamorphic rock (e.g., slate) that have neither inherent porosity nor permeability. Sedimentary rocks in the earth's subsurface may contain oil or water. The oil and/or water lend strength to the rocks called buoyancy pressure that can contribute to their integrity in space. However, when private and national oil companies drill wells to extract petroleum (with 8 barrels of brine for each barrel of crude oil) or pump water from aquifers, buoyancy pressure diminishes and the rock can lose strength. When pressure from the weight of overlying rocks overcomes the strength of rock in the subsurface, the subsurface rock compacts and causes the mass of rocks above them to subside slowly over time as fluid continues to be extracted. This results in a lowering or topographic depression of the land surface. Subsidence is a common happening worldwide where oil is pumped from the deep subsurface or water from aquifers. Mexico City obtains its water supply from aquifers and has subsided up to 8 meters in some districts, damaging structures and rupturing infrastructure. Houston, Texas does the same and has subsided to form a bowl-like depression that make the city very susceptible to flooding when tropical storms form the Gulf of Mexico blow in seasonally. Oil and gas extraction at the southern California Wilmington oil field caused subsidence of 9.5 meters between 1938 and 1958. The subsidence reached Long Beach City and its harbor and extended to Los Angeles Harbor. It cost more than $100 million dollars to repair damaged oil wells, pipelines, sunken wharves and harbor infrastructure, railroad tracks, streets and bridges. Engineers had to deal with the reversed flow of sewers and storm drains, and also design a program to arrest the subsidence. The subsidence was stopped by stabilization of the buoyancy pressure. The brine extracted with the oil was injected under pressure into the weakened subsurface sediments. This purely human activity that extracts water, oil, and natural gas from within the Earth does not physically endanger citizens but can be an economic drain on a community budget when remediation becomes necessary.

Wildfires

Wildfires occur naturally in localized areas but can extend to hundreds of thousands of acres. They most often occur when there has been an extended dry spell or drought. The fires can be started by human carelessness such as sparks from a campfire that is not totally out, or by live cigarette ends or pipe embers that are thrown onto dry terrain. Sparks thrown from wheels on train tracks as coaches roll by can ignite wildfires. Arsonists start wildfires. Wildfires are favored by low humidity and move quickly when driven by strong wind conditions. Warning systems have been set up in many communities that are at risk from wildfires because of proximity to vegetated areas that have ignited in the past. People who do not adhere to calls for evacuation die but the number of deaths is small. However, wildfires destroy the forests they rage through and any structures and infrastructure in their paths. They should not be great threats to the lives of expanded future populations but can affect peoples' livelihoods when tied in to silviculture and the continued development of forests resources: animal life, trees, and other vegetation.

Pollution

Pollution from human activities contaminates water sources and taints foods. This may be from industrial and agricultural operations that release toxins such as heavy metals and other inorganic chemicals, organic chemicals, and particulates into ecosystems waters or onto soils. Ecosystems are invaded as well by bacterial and viral pollutants from bad waste disposal practices from domestic and commercial sources, industrial and manufacturing facilities, and agricultural sites. Contaminated water cannot be used for drinking, cooking, hygiene or irrigation without the potential for dire health consequences for consumers from the ingestion of tainted foods or unsafe water.

Water supplies in developed countries are safe whether in urban centers or in rural areas where regulations on pollution capture and control are strong and enforced. This is not the case in many developing and less developed countries where sickness and death from pollutants can be rampant (e.g., heavy metal poisoning in China in 2008 and bacterial poisoning in Zimbabwe in 2008/2009). Nations that demographers predict will have the greatest expansion of populations (e.g., several in Africa, Asia, and Latin America) will need more water and more food. Citizens in countries with ineffective pollution control systems in place will be increasingly at health risk from pollutants in water, and soil.

Air pollutants include heavy metals as aerosols and particulates, and gases emitted mainly from coal-fired power plants, other coal-combustion industrial facilities, and smelters. The heavy metals (e.g., mercury - Hg) and particulates deposit on land and oceans via rainfall or gravity adding to the pollutant inventory originated from other sources. The results are the same: contaminated water and soil and threats to their usefulness as elements of food production for future generations.

Gas emissions pose another air pollution problem. Sulfur dioxide (SO_2) from coal combustion and smelting of sulfide ore minerals reacts in the atmosphere to solar radiation (sunlight) and with water (H_2O) droplets in clouds to form sulfuric acid (H_2SO_4), the principal component of ecosystem-damaging acid rain. Rain is naturally acidic because of the reaction of carbon dioxide (CO_2) in clouds with water to form carbonic acid (H_2CO_3). However, sulfuric acid is far stronger an acid and is the major contributor to acid rain. Acid rain as runoff in surface water can kill many life forms in lakes and rivers including food fish. This can affect a fresh water fish supply locally and diminish economic benefits from sport fishermen/fisherwomen and tourists.

Of more importance to national food supplies is acid rain on farmlands. The rain can leach major and minor essential nutrients from soils during rather short periods of time and lead to poorer quality crops with less yield per acre/hectare. Acid rain can also lay ruin to forests from its contact with the trees and from the leaching of nutrients from forest soils. Most developed countries have passed laws on capture and control of air pollutants that have greatly reduced the amounts of heavy metals, particulates, and SO_2 discharged into the atmosphere. However, some less developed and developing countries have passed similar laws but fail to enforce them. The result is that the fore mentioned toxic matter is still being emitted into the atmosphere and brought to the Earth's surface by gravity and precipitation to intrude global ecosystems.

This is especially true for Asia where Chinese emissions of Hg and SO_2 have been increasing as China continues its race towards greater industrialization. The Chinese Environmental Protection Agency has warned the country's leaders that acid rain has basically sterilized vast areas of what were fertile soils and is continuing to do so over similarly large tracts of farmland. This, added to arable land lost to urbanization and desertification, is threatening China's hard won capability to feed itself (1.35 billion citizens). Arable land has steadily declined in recent years from 127.6 million hectares in 2001 to 121.73 million hectares in 2007. The minimum farmland necessary to feed China's citizens is 120 million hectares if farmland loss ceases and the population stabilizes.[39] If it does not, China will suffer food deficits in the near future and will have to import comestibles. Also the continuing increase in Hg emissions from Chinese coal-burning power plants more than compensates for the marked reduction in Hg emissions from power plants in industrialized nations. Mercury is a neurotoxin that bio-accumulates in humans who breathe it and ingest it in foods, mainly fish. Mercury ingestion and bioaccumulation is especially dangerous to pregnant women and small children. This danger to their society can be greatly alleviated by Chinese government mandates but this has not been done effectively because of national economic aims. With the exception of Hg, air pollution problems are mainly localized or restricted regionally. However, the thinning of the ozone layer that is discussed in the next section is global in its reach and carries the threat of spreading skin cancer among important pockets of the earth's population.

The Ozone Layer - A Shield From Cancer-Causing Radiation: Degraded By Humans - Regenerating By Human Intervention

Life on earth has been protected from exposure to dangerous wavelengths of solar ultraviolet radiation called UV-C by the ozone layer. The ozone layer, at altitudes ranging from 15 to 35 km, filters out the UV-C rays. Man-made chlorofluorocarbons

(CFCs) released into the atmosphere from spray can propellants, from refrigerant in air conditioners for vehicles, homes, workplaces, and other locations, and from refrigerators, rose into and reacted with the ozone layer. The CFCs caused a thinning of the layer that subsequently developed holes over some geographic areas. This allowed the UV-C radiation to reach life on earth. Medical experts believe that exposure to UV-C rays is responsible for the high incidence of basal cell carcinoma (skin cancer) in Australia and other southern hemisphere countries. Global health experts predicted that continued degradation of the ozone layer would put many global populations at risk of contracting skin cancer or other maladies that might be caused by cell mutations triggered by the radiation. As a result, there was near 100% international collaboration to restrict the production and use of the CFCs by finding substitutes for them. This was done. It lessened the rate of ozone depletion and initiated a slow healing of the depleted ozone layer around the Earth that continues today.

The response of the global community to the threat of thinning and development of holes in the ozone layer was a seminal event. It demonstrated that countries worldwide could work together to protect today's population and future generations from global threats to their well being regardless of national political philosophies, religious and cultural beliefs, or economic status. In this case, the threat was skin cancer. One hundred and ninety-one (191) nations joined in a binding treaty to overcome this global threat to life.[40] We should note and emphasize that adherence to the treaty did not affect the economic development of countries significantly. This collaboration should be transferable to reduce the causes of what could be the greatest anthropogenic threat facing growing populations in the near future and generations thereafter: global warming/climate changes.

AFTERWORD

In the next chapter we will discuss the projected effects of global warming/climate change on Earth and its inhabitants if global warming continues unabated. However, unlike the minimal economic effects on nations signatory to the treaty to eliminate the sources of ozone depletion, attacking the sources of global warming with the same fervor will retard the economies of nations at all levels of development. The economic barriers to reach an agreement to reduce the driving force of global warming/climate change by cutting back on emissions of carbon dioxide (CO_2) and other greenhouse gases has to be overcome in the very near future. If adherence to a binding treaty that drastically cuts back on CO_2 emissions is not accepted internationally, all economies will suffer...all future generations will suffer... all life will suffer.

CHAPTER 6. GLOBAL WARMING/ CLIMATE CHANGE

Global warming is real. We see direct evidence of it in higher temperatures at global measuring sites and warmer ocean surface/near surface waters with increased moisture in the atmosphere over the oceans. It is evident in receding mountain glaciers, collapsing/calving and melting icecaps, and measurements that show that sea level is rising and that seawater has encroached onto land environments. Biologists have recorded the migration of animals (including aquatic species) and vegetation from ecosystems conditions changed by global warming to more hospitable environments. This leads to a loss of diversity in ecosystems and an interruption in a prey-predator food chain. There have been an increasing number of violent tropical storms, massive flooding, and extended droughts and heat waves in several regions. Many scientists worldwide have studied accumulated data (measurements and observations) and attribute these events in grand part to global warming.

The principal cause of global warming is the anthropogenic driven increases of carbon dioxide (CO_2) and other greenhouse gases (e.g., methane CH_4, nitrous oxide N_2O, chlorofluorocarbons CFCs), aerosols, and particulates in the earth's atmosphere. This global "barrier" mutes the complete transfer of heat absorbed by the Earth from the sun back into the atmosphere. Global warming is the result of the so-called greenhouse effect. Carbon dioxide is the gas most responsible for the global warming process, by both mass and volume. It makes up 72% of the total greenhouse gas emissions followed by CH_4 with 18%, N_2O with 9%, and 1% from other greenhouse gases.

It is important to remember, however, that CO_2 is an essential nutrient that feeds plant life. Plants use CO_2 in photosynthesis reactions and release oxygen (O_2) that is essential to animal life. Before industrialization-generated CO_2 began accumulating in the atmosphere, equilibrium existed between the CO_2 naturally released to the atmosphere and the CO_2 used in natural processes. Sources included CO_2 released by organic matter decay, respiration of life forms, volcanic activity, and other chemical reactions. Uses included CO_2 taken up by forest and other vegetation and CO_2 dissolved in ocean and other surface waters where it is used by life forms to build their calcium carbonate ($CaCO_3$) shells.

The industrial revolution began in the late 18th century in Great Britain, fueled by coal combustion. This burning of fossil fuel introduced large amounts of CO_2 into the atmosphere that had a natural concentration of 280 ppm (measured in dated ice cores). However, sinks (absorbing environments) for the gas in forests and surface waters were able to absorb and use the CO_2 thus maintaining equilibrium between emission of the gas and its subsequent uptake. As industrialization gained momentum worldwide, more CO_2 entered the atmosphere but sinks compensated for the increase and equilibrium was largely maintained. The successful drilling for and recovery of oil in 1859 led to its use as a fossil fuel for lighting and heating. The invention of the internal combustion engine followed. This invention and mass production and use of automobiles and airplanes further loaded increasing masses of CO_2 to the atmosphere. The concentration of CO_2 in the atmosphere today is 387 ppm (as measured in the atmosphere above Mauna Loa, Hawaii, 2008). This is 107 ppm or almost 35% increase above pre-industrial revolution CO_2 levels. Although small in actual concentration (0.0107%), the steady changes of increased CO_2 contents in the atmosphere catalyzed global warming and the climate changes that have ensued from it.

The content of CO_2 in the atmosphere has increased because of multiple factors. The world's increasing demand for power has brought about the construction of more and more facilities that burn fossil fuel. In 2008, China, was constructing a coal-burning power plant every one or two weeks to support the country's march towards industrial expansion and economic development. Coal supplies more than half the electricity-generating facilities globally, followed by oil and natural gas. The combustion of gasoline used by vehicles and jet fuel in airplanes contributes large masses of CO_2 to the atmosphere. As emerging economies improve with or without population increases, the number of vehicles on their roads increases dramatically as is the case in China and India. In a 2009 paper, The Energy Information Administration of the U.S. Department of Energy reported that the world's energy consumption is expected to grow by 44% during the next two decades (2010-2030). Overall, 94% of the world's projected increase in industrial energy use is expected to come from developing economies, with China, India, Brazil and Russia expected to account for 2/3 of the increase.[41]

As stated earlier in this section, there are two principal sinks for CO_2 in the atmosphere: 1) oceans and other water bodies; and 2) trees and other vegetation (terrestrial and aquatic). The sink capacity of the oceans is at its limits accounting for a portion of the CO_2 atmospheric increase. The sink capacity of trees and plants has been diminished by deforestation to harvest wood for construction and furniture, to make charcoal that is used in cooking, to clear areas for housing and infrastructure, for farming, and to provide range for livestock production. Much, if not most of this deforestation has been done in environmentally sensitive and important regions without enforced national mandates for reforestation. As global populations grow by about half or more during the next few generations, it is likely that additional vast vegetated areas (e.g., in the Amazon, in southeast Asia) will be lost as sinks for excess CO_2 in the atmosphere. The capacity of oceans and other water bodies, and trees and vegetation to take up CO_2 will be further strained as emissions from more sources of CO_2 discharge into the atmosphere. This will happen in spite of the good intentions of national laws, and non-binding regional and international treaties.

Increases in the rate of global warming will continue and bring about further climate change affecting larger areas of the earth. Yet, this need not happen if good will and good intentions expressed in international treaties bring about massive global investment with two aims. The first aim is to curb and reduce CO_2 emissions to the atmosphere using established technologies used now in the petroleum and chemical industries. The second aim is to sequester CO_2 captured by the technologies and either use the captured CO_2 or store it where it is permanently immobilized. A third aim is to adopt technology that can remove CO_2 from the atmosphere and capture it on sorbents from which it can be released and prepared for sequestration. These aspects for countering the global warming process will be discussed in the following chapter. If there is no immediate curbing and reduction of CO_2 emissions worldwide, many populations that inhabit our planet today and their children and grandchildren will suffer severe consequences that will diminish their quality of life.

It should be noted that CH_4, although present in the atmosphere in much smaller concentrations than CO_2, is a greenhouse gas that has a growing role in global warming. The CH_4 content in the atmosphere was 700 ppb (0.7 ppm) in 1750 and steadily increased to 1745 ppb (1.745 ppm) by 1998. Methane is important because its heat trapping capability is 20x that of CO_2 and its atmospheric concentration continues to increase.[42] Emissions from agricultural ventures (discussed later in the chapter), coal mining, oil and gas pipelines, wastewater plants, and landfills add to the CH_4 atmosphere's inventory. Indeed, in many areas, aged landfills are being tapped for their natural gas (CH_4) contents.

Effects of Global Warming

The effects of global warming on humankind today and on future populations as discussed in this section is derived in grand part from the International Panel on Climate Change report presented during 2007.[20]

Melting Ice and Rising Sea Level

Today's populations are experiencing the effects of global warming. Melting icecaps in the Arctic, in Greenland, and in the Antarctic, and melt water from mountain glaciers has caused a slow, progressive rise in sea level (20 cm [8"] in the past century). Land is being lost in coastal areas and the future survival of many island nations is at risk. There are scientific reports based on observations and measurements that demonstrate that the rate of global warming and glacial ice melt is increasing. If this continues, sea level is predicted to rise about a meter (>3 feet) by 2100. One foot of the rise will be from the expansion of seawater because of warming. A second foot rise will be from Greenland, the Arctic, and Antarctic glacial melt. The third foot will be from the melting of mountain (alpine) glaciers (e.g., in the Himalayas, the Alps, the Rocky Mountains, the Andes and elsewhere). People living within the three feet

(1 meter) elevation in coastal zones today number more than 100 million. A sea level rise of a meter would inundate their habitats and farmlands and displace this population. In Africa, rising sea level will inundate farmland and increase erosion in coastal regions of Egypt, Nigeria, Senegal, Gambia and southeast Africa, degrade inland ecosystems that yield essential natural resources, and displace populations. Bangladesh, Vietnam, Florida (USA), and many other coastal regions will see vast areas under seawater unless trillions of dollars are invested in sea walls to protect them. By 2100, perhaps 1/3 of the world's population (3+ billion people) will live within 62 miles (100 km) of a shoreline so that a one meter rise in sea level would likely result in a billion environmental refugees. An increasing rate of warming and melting would be disastrous. If the Greenland ice sheet were to totally melt over a period of time, sea level would rise about 7 meters (over 23+ feet). A melting of the West Antarctic ice sheet would give the oceans another 5 meter rise (16+ feet). In either case, seawater would increasingly inundate extensive zones of densely populated coastal regions and swallow up low-lying islands and island nations. This will slowly but surely disrupt social and economic order throughout the world.[43] Add to this the effects of storm surges and inland intrusion of sensitive freshwater ecosystems (see the following section) and the call for 100% global cooperation to slow, and even reduce the causes of global warming becomes a "save the world" call to action.

Intensified Storm Precipitation and Energy

In addition to causing an expansion of seawater volume, global warming increases evaporation so that the atmosphere over warmer seawater carries more moisture. This means that tropical storms will have increased masses of feedstock and will discharge greater volumes of water inland and cause flooding in susceptible areas. During short-term very heavy rainfall or as a result of longer-term extended rainfall, high volumes of inland rain will flow down stream to return to the ocean. In doing so, this water may flush through ecologically sensitive environments such as saltwater marshes and estuaries and dilute them, temporally disrupting ecosystems and harming the life forms they sustain. A rising sea level and more moisture to load into clouds also mean that storm surges will drive seawater farther inland. This will flood coastal zones, and will salinate fresh water environments including bogs and marshes, and agricultural fields that are within the reach of the surges. The seawater intrusions will also play havoc with ecosystems, temporally harm the life inhabiting them, and diminish the productivity of agricultural fields the seawater invades.

Drought

In contrast to the increased amounts of rain and tropical storm intensity are the extended drought conditions being experienced in many global regions. The less than normal amount of precipitation over time is cumulative and hurts agricultural projects. Regions in Australia, China, Argentina and areas in Africa, for example, experienced droughts in 2009 that have lasted over several years and that most scientists attribute to changing climate patterns caused by global warming. This affected livestock and crop production for domestic use in some areas and for export of these commodities. This reduced income for ranchers and farmers and export earnings from taxes for producer countries. Less availability of food in some countries as a result of drought coupled with war or political decisions prevent food aid from reaching needy populations. As a consequence, people suffer under-nourishment and starvation that is especially damaging to young children and the aged. Increased populations in regions that are likely to experience future long periods of drought will aggravate their food deficit situation.

Reduced Water Supply From Receding Mountain Glaciers

Receding mountain (alpine) glaciers present a grave problem for more than 1.2 billion people who depend on seasonal melting after a winter buildup of snow and glacial ice to renew their water supplies. With global warming glacial melt will yield high peak flows that will be followed by a sharp reduction in river flow and in freshwater availability as glaciers recede. The decrease of water supply to rivers and aquifers and its ultimate loss from mountain glacier melt portends badly for people dependent on this water supply and for their way of life. This result of global warming will affect populations on the Indian sub-continent, Europe, Asia and countries in South America. Except for Europe, populations in these areas will continue to expand for a few generations or more. This will put great stress on governments to provide basic water supplies for drinking, irrigation, cooking and hygiene in 40 to 50 years that they struggled to supply in 2009.

Climate Change and Spread of Tropical Disease

Shifting climatic zones in tropical/sub-tropical regions caused by global warming can extend the reach of diseases. Malaria has broken out in once-cooler areas where malaria was not known such as in the highlands of Papua, New Guinea, Ethiopia, Tanzania, Kenya, Rwanda, Burundi, and in Cambodia. In New Guinea, for example, the highlands at altitudes greater than 1800 m did not provide an environment for mosquito breeding. The fact that these areas now have mosquitoes that infect people with the malaria parasite is thought to be the result of two factors that have been measured by scientists in the field: warmer temperatures and more rainfall. The rainfall pools up and provides breeding sites for mosquitoes and warmer temperature abet a more rapid breeding. In warmer and moister climates caused by global warming, other vector borne diseases transmitted by insects and animals expand their ranges and will cause increased numbers of health problems for society. These are illnesses such as dengue fever, yellow fever, viral encephalitis (several variants), sleeping sickness (tsetse fly), and others. Bugs (e.g., bark beetles) and fungi that kill trees and other vegetation will also expand their ranges and harm ecosystems. Also, more CO_2 in the atmosphere encourages growth of ragweed and concentration of tree pollen thereby lengthening allergy and asthma seasons. These conditions can also give rise to increases in food-borne and water-borne diseases such as cholera, salmonella, e-bola, and hepatitis. The tropical/sub-tropical maladies that have spread into environments where they were not previously known have infected today's populations with unsettling effects. Many of the newly affected areas are precisely where population growth is estimated to be high for at least the next 40 or so years. The larger populations will be susceptible to the continued spread of diseases because of global warming until methodologies and therapies are found to protect people from them (e.g., vaccines) and to control and eliminate disease as essentially has been the case for smallpox and polio.

During 2008, the University College London Institute for Global Health and the British medical journal The Lancet created a commission to evaluate the effects of climate change on the health and well being of an expanding global population and present recommendations for confronting projected health problems. The Lancet Commission, headed by Dr. Anthony Costello, assembled a group that included engineers, political scientists, lawyers, geographers, environmental scientists, anthropologists, economists, development planers, philosophers and students, in addition to health scientists and practitioners, In a 2009 paper titled "Managing The Effects Of Climate Change" in The Lancet, Dr. Costello and his collaborators

assessed the general major direct and indirect threats to global health (and personal security - author's addition) from climate change through changing patterns of disease.[44] These
were divided into four categories: 1) from population growth
and migration; 2) from water and food security; 3) from extreme climatic events; and 4) from vulnerable shelter and
human settlements, themes discussed earlier in this book.
Dr. Costello and his collaborators correctly emphasize that these
are likely to have the biggest effect on global health. They duly
re-emphasize that the effects on health from climate change
will worsen the disparities between rich and poor segments of
societies in parts of the world where they exist now, particularly
in less developed and developing nations.

Climate Change and Agriculture

Weather is the principal control on agricultural productivity. Temperature has a direct influence on rainfall, glacial runoff,
and is responsive to the amount and duration of sunshine (or
lack of it). Together, these factors and the nutrient capability of
soils affect the capacity of farmland to produce food farm crops
for humans and fodder for food animals. The primary factors
in climate change as they affect today's world are temperature
increase from global warming and its geographic distribution.
In addition to its influence on precipitation and runoff, temperature controls transpiration from vegetation and moisture
stored in soils. These meld with nutrient contents to support
soil fertility.

In its own way, agriculture contributes significantly to
greenhouse gases in the atmosphere and agriculturalists are
working towards minimizing farm output of greenhouse gases.
For example, in the United States, agricultural activities produce and release more than 16% of greenhouse gases emissions:

CO_2, CH_4, and N_2O. Globally, the figure is 18%. The CO_2 increase comes in part from clearing forests and burning crop residue; N_2O is from fertilizing with nitrogen-rich chemicals; and the CH_4 emanates from submerging land in rice paddies and raising large livestock herds via their burping, flatulence, and decaying wastes.

Scientists in India calculate that the country's 283 million cattle released 11.5 million metric tons of CH_4 to the atmosphere in 2007 from enteric fermentation expressed as burping and flatulence. Extrapolating this to the 1.4+ billion cattle globally gives an annual CH_4 gas release to the atmosphere of more than 57 million metric tons from cattle. The demand for more milk and other cattle products by larger numbers of people on earth and by global levels of increasing affluence will require more cattle and drive up livestock CH_4 contribution to the atmosphere and abet global warming. Add to this the contributions from piggeries and herds of sheep, goats and other ruminants worldwide and it is clear that this mass of CH_4 gas should be reduced where possible. Indian scientists are studying how changes in feed and other conditions can be modified to diminish the cattle CH_4 output. Some are investigating the potential for CH_4 capture and use.

Canadian scientific teams have used traditional techniques to breed cows that produce 24-28% less CH_4 than cows not bred with the same purpose. They do this by examining genes responsible for fermentation and CH_4 produced by a cow's four stomachs and then finding animals with the diagnostic genes for the breeding. The scientists also found that when farmers feed livestock a diet higher in energy and edible oils, there is less fermentation than in livestock fed grass or lesser quality fodder. Farmers in Canada using this technique accrue carbon credits of $1-10 Canadian per head. In New Hampshire, a few dairy farms reduced CH_4 emissions from their cows by an average of 12% by using a livestock feed of alfalfa, flax, and hemp. Clearly, science and agriculture are on the correct path in applying methods to limit the agricultural contribution of CH_4 to global warming.[45, 46]

Latitudinal Shifts of Agricultural Zones

Global warming has given rise to higher temperatures towards the poles in middle and higher latitudes. The result is a shift in ranges of plant and animal species towards the poles and to higher altitudes that causes a change in agricultural production patterns. For principal crops in some regions, increasing atmospheric CO_2 levels that drive global warming are beneficial to agricultural productivity. In others regions, crop productivity is reduced. For example, there is a physiological response of crops to increased CO_2 in the atmosphere. Wheat, rice, and soybeans have higher productivity with increased CO_2 levels, whereas corn, sorghum, sugarcane, and millet do not and have declines in productivity. The warming has a positive effect of extending the length of a growing season for some agricultural zones. In temperate latitudes of the United States, for example, the growing season has extended 14 days over a 19-year period. This allows earlier planting that gives rise to a sooner greening of vegetation in the spring and earlier harvesting of crops. Extended growing seasons present the possibility of growing two crops. On the negative side for agriculture, warmer temperature allows for weed creep into higher latitudes and higher altitudes. Also, higher temperature can hasten crop maturation (leaf unfolding) and result in reduced crop yields. In the United States, corn and soybean yields fall by 17% for each degree increase in the growing season average temperature. Higher temperatures favor the migratory invasion of a wider variety of pest insects and their proliferation particularly in warmer temperate climates. This could lead to an increase in outbreaks of crop diseases. Countering these pest/disease problems that can reduce crop yield and crop quality (nutritional value) requires proactive weed and pest insect and animal management programs.

During the last century, average global temperatures have increased 0.76°C (1.36°F). If all input of greenhouse gas emissions had ceased in 2000, global warming momentum would have increased the world average temperature another 0.6°C (1.08°F) by 2050.[20] The overall effect of a moderate climate change (1-2°C) on world food production may be small because the temperature rise is not the same worldwide. Thus, reduced production in one region is likely to be balanced by increased production in another region. Crop productivity is projected to increase slightly at middle to high latitudes for local mean temperature increases of 1-3°C depending on the specific crop. High latitude crop producing countries such as Canada and Russia may develop larger cropping areas northward but crop yields will probably fall short of expectation because of less fertile (podzolic) soils there. For seasonally dry and tropical regions at lower latitudes, the forecast is for decreased crop yields even for 1-2°C (1.8-3.6°F) increases. Globally, the projection is that potential food production will increase over the range of 1-3°C (1.8-5.4°F) but decrease if temperatures rise above this range.

Dr. Costello and his research group see the threat to food security as being triggered when a safe threshold they set at an average 2°C (3.6°F) above the pre-industrial age average is exceeded, with the realization that the temperature increase will be larger at higher latitudes. The threat becomes a medium risk if the temperature advances 2-3°C (3.6-5.4°F) by 2090 or a major risk with a 4-5°C (7.2-9°F) build up in northern Canada, Greenland, and Siberia.[44]

Some Regional Consequences of Climate Change

The effects of climate change on the food supply in many less developed and developing nations that are most susceptible to a diminished supply is a of heightened international concern. Many of these nations will experience great increases in their populations by 2050. In Africa, less water will likely be available for irrigation and desertification will undoubtedly worsen because of a reduction in average precipitation due to global warming. This is expected to be felt most in southern, northern and western African regions and will cause increased water stress for up to 250 million Africans by 2020. Agricultural production (e.g., grains) will be threatened by both the global warming temperature rise and variability of rainfall. By 2020, yields from rain-fed agriculture could fall by 50% in some countries. Soil salination will increase. Especially at risk from these conditions are arable areas that could shrink, marginal arid and semi-arid areas that could suffer a drop in crop yields, and agricultural zones with contraction of their growing seasons. In order to moderate the effects of temperature increases, reduction and variability of rainfall, and the consequences they bring to bear, countries will have to adapt to the climate change effects. Africa and its 53 independent states is least likely to be able to adapt for several reasons, prominent among them the lack of economic resources. There will be a loss of food security in much of Africa. Adaptation is the subject of the final chapter of this book.

In East and Southeast Asia, crop yields by 2050 could increase up to 20% but they could decrease up to 30% in central and south Asia. The hunger risk in developing nations in the region will be high because of growing populations and urbanization at the expense of agriculture.

The forecast is that agricultural (and forestry) production over much of southern and eastern Australia and parts of eastern New Zealand will decrease by 2040 because of warming. The warming in these regions adds to periods of drought and sets the stage for wildfires. Australia has been plagued by drought for the past several years that led to an 80% drop in rice production. However, in areas of eastern and southern New Zealand near major rivers, agriculture production is expected to increase because of warming that extends the growing season, provides more rainfall and results in less frost. During 2009, Australia suffered massive destructive wildfires that were caused in part by arson and carelessness and abetted by low humidity, dryness of vegetation, and high winds.

Southern Europe will suffer a decrease in agricultural production as a result of a warming shift to the north. However, northern Europe will benefit from the temperature rise in higher latitudes by increased crop yields (and forest growth).

In drier zones of Latin America, an increase in temperature by 2050 will hurt agriculture projects because of loss of soil moisture, salination, and in some cases desertification. Warmer temperatures are expected to cause a decrease in productivity of some important crops. Livestock productivity is also likely to decline.

North America is projected to have increases in aggregate yields of rain-fed agriculture by 5-20% because of moderate climate change during the first few decades of this century.

However, crops near the warm end of their suitable range that depend on high water use will have problems in maintaining their agricultural yields.

Rising Global Temperatures and Aquaculture

Food fish provide valuable sources of protein for hundreds of millions of people worldwide, whether in coastal zones, in inland lake and/or river regions, or in urban centers. Warming of ocean waters has driven shifts of algal and plankton distribution and a consequent migration of some fish species to more hospitable cooler waters. This has hurt near shore fishing that supplies the food/protein needs of innumerable coastal villages in Africa and in Southeast Asia. Small islands will experience reduced fisheries yields. Migration of prey species has been followed by migration of predator species. This affects ocean fisheries. South America will experience shifts in locations of southeast Pacific fish stocks over one or two generations because of an increase in sea surface temperature (abetted by sea level rise) off its west coast. A shift in fish abundance in high latitude areas is attributed in part to migration of nutrient (algae, plankton, prey) sources away from warmer ocean waters. There have also been increases in algal and zooplankton abundance in high latitude lakes. Observed and measured shifts in fresh water fish species correspond with changes towards higher water temperatures and nutrient source migration in lake and river fisheries. In Africa, for example, local food security may be threatened in the near future by decreasing fisheries resources in large lakes because of rising water temperature. Range changes and earlier migration of riverine fish has altered fisheries programs. Global warming affects freshwater and marine aquaculture fish production as well because of regional changes in distribution and production of particular fish species. As noted previously, aquaculture supplies about 40% of the food fish demands of our global population. Thus, aquaculture projects have to modify their products and methodologies to deal with changes caused by global warming. Only in this way can aquaculture maintain or improve production levels and provide for increases in populations in nations where aquaculture is a major supplier of protein-rich food.

Heat Waves

Global warming will likely cause higher temperatures during heat waves and increase how often they occur and how long they last. Scientists studying global warming estimate that by 2100 there will be a quarter more heat waves in Chicago and a 4- to 8-fold increase in Los Angeles. Because of this and expanded global populations, particularly in densely populated urban areas, heat waves could kill high numbers of people unless governments prepare to protect their citizens against them. In a previous section we learned that the city of Chicago put a plan in place that cut the number of deaths from a heat wave in 1995 from 739 to 103 during a heat wave four years later.[47] This action is commendable. However, other areas worldwide did not have heat wave action plans and suffered accordingly.

In August 2003, a heat wave in Europe caused 34,000 heat-induced deaths, more than 14,000 of them in France. The summer of 2003 was the hottest summer in Europe in 500 years. In Germany, the temperature reached more than 104°F (40°C) and in England the high temperature was 100.6°F (38°C). In Portugal, 18 people died from wildfires abetted by the heat wave that killed 2100 people and destroyed 740,000 acres of forests and 110,000 acres of farmland. To ameliorate their national problem, the French government passed a National Heat Wave Response Plan that allocated $6.8 billion over 5 years to prevent massive heat-induced fatalities. Other countries have initiated their own action plans.[48]

Low elevation non-coastal central California suffered through a deadly heat wave from mid- to late-July 2006 that caused the seven hottest days ever registered for the region. Temperatures peaked from 22-26 July with triple digit record daytime readings up to 113°F (45°C) and record overnight temperatures up to 90°F (32°C) well above the previous maximum of 83°F (28°C). A combination of high humidity and heat prevented the normal overnight cooling. The high humidity is thought to be in part the result of global warming that increased temperatures of ocean waters. Although rare for the normally dry conditions in the Central Valley, high humidity has become more common in occurrence since the 1990s. The heat wave was blamed for 147 deaths but this number may be under reported because of multi-factors that can bring on heat-related mortality. In addition 25,000 cattle died and crops were lost.[49]

In 2008 and 2009, Australia was hit hard by heat waves, wildfires, and drought. A heat wave in South Australia during January 2008 had temperatures of 95°F (35°C) for 15 consecutive days and in March temperatures were above 100°F (>37°C) for 7 consecutive days. During 2009, a bushfire in Victoria, possibly started by spontaneous combustion, killed 173 people and destroyed 2029 homes.[50, 51]

We cannot prevent wildfires/bushfires started by natural means aided by high temperatures such as spontaneous combustion or those started by lightening strikes. However, we can establish action plans and conditions to prevent unnecessary deaths by heat waves if governments have the strength of purpose and economic resources to do so.

AFTERWORD

In past chapters we discussed the sustenance and ecosystem issues affecting growing populations today and that expanding populations will face in the next few generations (40-50 years). We considered the stresses put on populations and the ecosystems they inhabit as well as existing problems and future ones governments will encounter. These are to provide enough safe water and untainted food to sustain citizens and natural resources to support economies. We presented the positive and negative aspects of what could be done to meet the basic needs of populations today and in the future. We reviewed natural biological and physical events and anthropogenic-originated happenings that have ravaged populations in the past and those that affect people today. We considered how to cope with failings in dealing with such disasters by responding to them efficiently and effectively based on what evaluation teams learned from their studies of past events that could be applied towards future planning. We can use this knowledge to moderate or even eliminate many risks that natural and anthropogenic disasters pose to people and ecosystems. As discussed in this chapter, global warming/climate change and the effects on populations throughout the Earth and as projected for the future is a great immediate concern to the international community.

We will continue the discussion of global warming/climate change in the next chapter following two tracks. First is how humans can adapt to climate change. Second is what could be done (conceptually) to moderate, stabilize, and possibly halt or reverse the rate of global warming beyond cutting back severely on the emissions of CO_2 and other greenhouse gases to the atmosphere.

CHAPTER 7. ADAPT AND INNOVATE: LOGICAL + OUT-OF-THE-BOX BRAINWORK COUNTER EFFECTS OF CLIMATE CHANGE + LOWER CO_2/BLOCK SUN

ADAPTATION

Adaptation is a long-term process during which a population of life forms adjusts to changes in its habitat and surrounding environments. Adaptation by the world community as a whole is vital to cope with the effects of global warming and climate change that serve to rupture the nutrient-energy continuum in ecosystems.

The global community must act now to adapt to changing environmental conditions. This will increase human capability to adjust to gradually changing future effects of global warming/climate change that can alter the quality of life for all populations worldwide. Among the most important of these effects are shifts in agricultural zones and changes in agricultural productivity, reduced water supply (per capita), rising sea level, and the spread of tropical diseases. Civilization can adapt to global warming induced climate change. This can be accomplished by integrating engineering and technology, bioscience, ecology, and proactive agricultural management programs into an overall design to minimize the evolving impacts of global warming and ensuing climate change on the well being of all life on earth.

Farming

Farmers can adapt in order to maintain or improve productivity in several ways. They can change seasonal patterns and alter planting times (sow seed earlier), switch to later-maturing crops or species, modify cropping sequences, and plant different varieties or species that can thrive under the changed climatic conditions. Farmers can conserve moisture through better tillage, by improving water supply, and by employing efficient irrigation methods (e.g., drip/focused irrigation). The use of optimal amounts of fertilizer and other agricultural chemicals to sustain growth and reduce runoff will prevent waterway pollution from excessive nutrients and also will cut back on farming expenses. In India, farmers in some agricultural zones facing increased flooding that is likely a result of climate change are starting their crops in nurseries to plant them after flood waters recede. They are also growing crops in extensive planters constructed above flood level and when floodwaters subside, grow crops beneath the planters. Finally, farmers can use grain drying and storage that minimizes loss to rodents, insects, fungi or bacteria. Bio-scientists can work towards sustainable agriculture by breeding new crop varieties (altered cultivars) that are resistant to disease, pests, heat, water stress, and drought. To do this they can use traditional hybridization and marker-assisted selection hybridization for same species breeding, and genetic engineering for cross-species breeding. The European Union group and other countries do not accept the latter.

Rising Sea Level

The approaches to dealing with rising sea level are multi-fold. The first is to slow down or arrest the ice sheet and alpine glacier melting by alleviating or eliminating the driving forces for global warming. This is a long-term effort that requires the halting of input of additional CO_2 and other greenhouse gases to our atmosphere and ultimately reducing their masses in the atmosphere. The reduction of additional greenhouse gas release into the atmosphere can be accomplished by today's technologies and can be effective if governments enforce regulations that require industries to do so. The cost to industry and economic development will be high but will help ameliorate a global problem that puts the emitters' future existence in question.

A second relatively shorter-term effort is to design and put in place sea defenses appropriate to a threatened coastline such as fixed walls high enough to protect people and property from a projected 50 or 100 year future sea level rise and from surges of ocean water inland during high energy storms. Sea wall defenses can also be hydraulically driven electrical systems with solid walls that can be lifted to increasing heights in response to a sea level rise and storm surges. However, the global costs to do this would be in the trillions of dollars, euros, or other currencies and would put such projects beyond the economic capability of most nations. An option would be to move coastal populations, cities, infrastructure, industries, and agricultural production inland out of the reach of an encroaching sea. This is really not actionable given the magnitude of such a move. Humankind has to provide the economic resources to allow engineers and global technological experience and capability to design and put in place defenses against sea level rise at all populated locations where they are needed.

As noted above, sea level rise and encroachment defenses have the added benefit of moderating the effect of surges onto inland areas from high-energy tropical storms. Many meteorologists believe that such storms are increasing in number and intensity as a result of global warming. Extensive wetlands sap energy from tropical storms as they pass over them on their path inland. Vast areas of wetlands have been destroyed by human activity for economic benefit. These wetlands have to be reclaimed to reduce the threat to population centers from hurricanes (typhoons, monsoons). The loss of life and property in New Orleans and the southeast coast of the United States as a result of Hurricane Katrina would likely have been significantly lessened if the coastal wetlands ecosystems had been in their natural, undisturbed states.

Reduced Water Supply

Adaptation to reduced water availability from receding glaciers that could disappear in the future and from irregular variability in timing and amount of rainfall a region receives are problematic today. They will be serious problems in the generational future. Planning today is essential to cope with this water problem that exists now in some regions and will worsen with time. As discussed previously in this book, adaptation to a reduced water supply includes several tactics: 1) hybridization to develop food crop species that tolerate reduced rain or drought; 2) piping water from where it is plentiful to where it is in deficit - canals and aqueducts can be used and can preserve the water supply if they are covered to eliminate loss by evaporation; 3) recycling domestic and industrial waste water; 4) developing new sources of water (e.g., build desalination plants, find new aquifers, extract water from the atmosphere) where the commodity is insufficient to sustain the needs of existing and growing populations; and 5) store water when it is plentiful to deal with long-term deficits.

Health Issues

Adaptation by populations to organic (biological) and physical health threats from global warming and climate change is essential to minimize their impacts on people and the ecosystems that sustain them. As we learned in earlier chapters, these effects are direct and indirect. The direct health effects that can be attributed to global warming/climate change are the changing patterns and spread of infectious (contagious) diseases (e.g., cholera, typhoid, influenza) and non-communicable diseases (e.g., cardiovascular disease, diarrhea, malaria, dengue fever). They also include sickness and injuries cause by extreme heat, high-energy storms, increased flooding and mass movement (e.g., landslides), and drought. Other direct health effects come from a lack of food security (e.g., malnutrition, starvation), food safety and contaminated water (e.g., diarrhea, cholera), vector-borne and rodent-borne diseases, and other infectious diseases. A major indirect effect that affects the well being of a population is the lack of finances to deal with health problems in under-resourced countries most affected by global warming/climate change. The availability of funds will allow the affected nations to educate and train human resources and to create or strengthen the infrastructure that supports health services.

Adaptation to cope with the effects of global warming on the spread of disease and disease control falls into four general categories. To achieve success for each requires the input of massive amounts of investment by developed economies. By helping less economically advantaged countries to adapt and be able to manage contemporary health problems from global warming and climate change and those that will spread and manifest themselves in the future, the developed nations help safeguard their own futures.

One adaptation to protect populations from disease or to prevent the spread of infectious diseases is the development of vaccines. Major pharmaceutical companies and many small ones are researching formulations for vaccines for a broad spectrum of diseases because they project that this is an area in which substantial profit exists. The diseases include Alzheimer's, AIDS, genital herpes, urinary track infections, grass allergies, pandemic flu variants, meningitis, and travelers' diarrhea. The delivery methods vary from shots and nasal sprays to patches and pills. The expectation is that many of these vaccines and other delivery methods will be ready to protect populations in 5 years or less. As these become available, they will join the successful ones that have been formulated and are used globally against diseases such as smallpox, polio, German measles, seasonal influenza, ear infections, pneumonia, blood infections, rotovirus, cervical cancer, and seasonal influenza. The markets for vaccines are worldwide and growing, both from awareness and expanding populations. Many vaccines and pharmaceutical therapies require annual applications. In 1990, for example, 20 million people received seasonal flu shots in the United States and by 2008 the number had increased to 113 million. The sheer volume of vaccinations at reasonable costs will increase profits for pharmaceutical companies.

In the case of a disease that is a variant of a group for which vaccines have been prepared in the past, as for influenza, it is possible to prepare a vaccine fairly rapidly. This was the situation for the swine influenza, the A H1N1 virus, in 2009. The time between the research to create the A H1N1 vaccine, its successful (albeit shortened) clinical trials, and its initial availability to health clinics took only about 6 months. This was too late for the first waves of those who contracted the A H1N1 influenza strain in early 2009 but served to protect populations as the pandemic progressed during the last quarter of 2009 and during 2010.

In other vaccine development programs, scientists can spend years without formulating an effective vaccine. This is the situation for dengue fever, a malady that is expected to spread with global warming and climate change to 6 billion people by 2050. Nonetheless, scientists keep working on vaccines with the hope that there will be a breakthrough. This might be the case for vaccines that have been created and tested for more than 25 years with the aim of protection against AIDS, an infectious disease not related to global warming/climate change. None were successful. However, recently an experimental AIDS vaccine was formulated from two failed vaccines to work against two subtypes of the AIDS virus. It protected more than 30% of people from the AIDS virus variants identified in Thailand and other areas of southeast Asia, and in the United States and Europe, giving hope that a truly effect vaccine would eventually be developed.[52] Substantial funding will be needed to support high level research into vaccines against infectious diseases and non-communicable diseases that can impact populations worldwide.

Another adaptation is to further research to improve existing therapies or develop better ones against those diseases expected to spread to higher latitudes and higher altitudes (changed distribution of some disease vectors) as a result of global warming and climate change (e.g., malaria, dengue fever, tick-borne encephalitis). There are other diseases that will increase in geographic reach and in the numbers of citizens affected as a result of global warming/climate change. They must also be foci of intensified research to discover how to eliminate them or improve therapies for them. These include schistosomiasis (disease caused by elongated trematode worms that parasitize blood vessels and tissue), fascioiasis (disease caused by liver flukes), alveolar echinococcis (disease caused by a small tapeworm), leishmaniasis (disease caused by leishmanias [flagellate protozoans - parasites]), Lyme borreliosis (a tick-borne disease, especially deer ticks), hantavirus (a rodent-borne disease, especially deermouse), and arbovirus (viruses transmitted by arthopods including causative agents of encephalitis, yellow fever, and dengue fever).[44]

A third adaptation is to establish public health services at locations where they are lacking and to improve existing health delivery programs. This requires economic investment to develop a health system's infrastructure and personnel. The personnel include medical doctors who can diagnose problems and prescribe the proper therapy, physician assistants/nurses, and technical and service staff. The infrastructure includes clinics with diagnostic tools, medical supplies, pharmaceuticals, and transportation/delivery support that can respond effectively to health threats whether or not they are driven by global warming/climate change. Where existing public health services are fragmented and unconnected among urban and rural components, they must be brought under one directive umbrella to achieve a ready exchange of information that improves public health services. This includes educating the public on disease prevention. Public health services will bear the bulwark of the higher incidence of infectious and non-communicable diseases that will propagate with expanding populations in an era of progressive global warming and ensuing climate change. This will be most pronounced in less developed and developing nations.

The economic input to achieve adaptation and cope with health issues, localized or global, will have to come mainly from developed nations. Substantial help should come as well from emerging economies that have accumulated large foreign exchange reserves. This can originate by taxation of those major industries (now, in many cases, multinational companies) that prospered in times of little or no environmental stewardship.

It may also be from an international transportation tax or sales tax, by added fees on export/import commodities or a combination of these options. The industries have a moral and ethical obligation because their unfettered economic progress brought the Earth to a precarious environmental stage. Financing that reaches nations in need of health service rebuilding has to be monitored closely so that funds are used where they are needed and not lost to corruption that is rampant in many of the most needy countries.

Barriers to Adaptation

Barriers to implementing adaptation of methods and technology to address changes that global warming imposes on populations of our planet are many and varied. They require global negotiations and legally binding treaties among all nations to mitigate negative effects and to take advantage of positive effects that people will experience in different global regions.

The nature of impediments to meeting the challenges of warmer temperatures and shifts in climatic zones that affect the quality of life for growing populations fall into several familiar categories. One is informational: will there be an open transfer of methodologies and technologies from industrial societies to less developed and developing nations in the efforts to counter the effects of global warming? A second is economic: are there financial resources to carry out necessary projects? Another is environmental: will changes that moderate one eco-problem cause eco-problems elsewhere? A fourth is social: are our global populations willing to make necessary sacrifices in their life styles in order to combat the effects of the warming on the Earth and its resources? Finally, there is a cultural and attitudinal obstacle: can religious/cultural beliefs and behavioral and attitudinal norms be moderated to accept changes that can reduce global warming and preserve or better the way of life for believers and other people on Earth? In light of the global chaos that an unfettered warming of the planet will cause if left to increase unchecked, the nations of the world, whatever their politics, will surely come together to fight global warming. The question Mother Earth has is: will nations respond now or act contrarily and put off what they have to do to contribute to coping with the problem because of their own economic/political/social aims? Indecision to act allows options for attacking this global problem to evaporate.

Adaptation Couplings: Mitigation and Development

The United Nations Framework Convention on Climate Change emphasizes that for adaptation to global warming/climate change to have positive results, it must go hand-in-hand with mitigation actions that reduce CO_2 and other greenhouse gases emitted to the atmosphere. As developed (industrial) nations cut back on emissions of CO_2 (and other greenhouse gases) to lessen the rate of global warming, less developed and developing nations have to implement action plans correspondingly. This has not been the case. During 2007, China emitted more CO_2 of the world total in 2006 (21%) than the United States (20%). China's contribution together with those from India and Russia (31%) exceeded those of the United States and Japan (26%).[53, 54] As the developed (industrial) nations cut back on their CO_2 emissions, those from developing economies continue to expand. This reality has to change and means that adaptation and development have to work together in the mitigation process. Thus, governments have to instigate and fully implement national development programs that have adaptation as a prime priority. The programs must be supported by development projects that are based on mitigation technologies, directed by a nation's fiscal condition, and supported by international financial and technical aid. During November 2009, the European Union committed 70 billion euros over a 10-year period (through 2020) to that end. Other industrial (developed) nations and international organizations have been granting funds for combating the effects of global warming/climate change on agriculture, water, health, coastal management, and other sectors. This economic assistance carrying with fiscal it transparency requirements can be expected to increase substantially within the 2020 timeline.

MODERATING/HALTING/REVERSING GLOBAL WARMING

There are two main tracks that can be followed in order to reduce or even halt global warming. The obvious track has been discussed several places in this book: reduce the mass of CO_2 and of other greenhouse gases (CH_4, CFCs, NO_x) that are released into the atmosphere. This is coupled with improving the "sink" capacity of the Earth's terrestrial and marine ecosystems and would complement the "at source" CO_2 capture, liquefy, and sequester technologies now in practice or being researched. The less obvious track is to thoroughly evaluate the geo-engineering potential for preventing a portion of incoming solar radiation from reaching the earth by blocking it and reflecting it back into the stratosphere.

Fertilize the Oceans - Allow for Greater Uptake of CO_2

This is basically an adaptation of the reforestation concept to increase the Earth's sink capacity of vegetation for CO_2 to help mitigate the CO_2 increase in the atmosphere. The concept is that the oceans and seas can be fertilized with iron, phosphorus, or nitrogen in order to stimulate growth of plankton so that greatly increased masses of plankton will take up CO_2 for photosynthesis (vegetation) or to build their carapaces (animals). Upon dying these organisms carry the carbon-bearing tissues to the seafloor. A potential environmental problem is that this will upset the balance between organic matter deposited on the ocean bottom and the demand for biological oxygen that fuels the decomposition of the organic debris. If more organic matter is deposited in the deep marine environment than can be decomposed by the oxygen there, eutrophication will set in as has been the case in the Baltic Sea. The lack of oxygen in bottom waters will establish an ecosystem that will not support most ocean life. Although conceptually interesting, this method of extracting CO_2 from the atmosphere is thought by many to be environmentally flawed.

Extract CO_2 From The Atmosphere

The technology exists to take nitrogen (N_2) out of the atmosphere and to extract water from the atmosphere. Surely, CO_2 can be extracted from the atmosphere using engineering techniques learned from N_2 and H_2O extraction methodologies. Clearly, the extraction of N_2 from the atmosphere where it has a concentration of 79% has been economically actionable. The extraction of CO_2 with a 0.0387% concentration can be done technically. In the simplest approach, envision a simple pump that sends air flowing into a calcium oxide (CaO) solution with a pH value 8.1. The CO_2 in the airflow into the solution will precipitate out as calcite or aragonite, mineral polymorphs of calcium carbonate ($CaCO_3$), the minerals from which seashells are built. Obviously, given the low CO_2 concentration in the atmosphere, the extraction idea initially seems far-fetched. However, when we ask "is it doable?", the answer is yes. It can be an important adjunct to reduce CO_2 in the atmosphere when coupled with a reduction of emissions from anthropogenic activities. Is the concept feasible economically? When confronted with what can happen to national economies and population well being in a wait and see scenario, the answer is yes.

A first stage in an atmospheric CO_2 reduction project would be to achieve concentration stability even as less developed and developing countries continue to emit increasing masses of CO_2 to the atmosphere. A second would be to reduce CO_2 concentration in the atmosphere as all countries cut back greatly on their emissions. This will be a long-term project because of the small CO_2 contents in the atmosphere. To be successful, this project would require the cooperation of all nations on the planet. The long-term benefits for humankind would far exceed the costs in technology and labor to carry out the project.

It took more than 200 years to bring the CO_2 in the pristine atmosphere at 280 ppm to a 2009 level of 387 ppm. It will take time to reverse the trend by decreasing CO_2 input to the atmosphere and increasing the rate of CO_2 extraction from it to reduce the atmospheric CO_2 contents. Scientists calculate that lowering the atmospheric CO_2 level to at least 325 ppm will first stabilize and then reduce the rate of global warming at a tolerable but not optimal level for the Earth's future increased population.

Scientists researching this concept have to make several determinations as they work out a benefit/cost number that will affect the feasibility of implementation of the proposed plan. They must first calculate how many pumps (hundreds of thousands, millions?) of a yet undetermined capacity to draw in air will have to be in operation. Second, they must establish where geographically the facilities for pump-precipitate operation should be located for maximum efficiency and effectiveness.

Third, they have to determine how long the systems will have to operate to bring about stabilization and then a decrease in the atmosphere's CO_2 concentration as part of an overall CO_2 "reduction at sources" package. The answers to these fundamental questions are predicated on whether the rate of greenhouse gases emissions to the atmosphere is assumed to continue increasing, to stabilize, or to decrease.

Block Part of Solar Radiation From the Planet

Another approach towards reducing or even reversing global warming is to reduce the amount of solar radiation reaching the Earth by blocking it and reflecting it back into the atmosphere. A reduction of 1-2% would stop the increase in global warming.

Volcanic Eruptions as Global Cooling Events

Historical and recent volcanic events give scientists insight on how the Earth can be shielded in part from solar radiation. The Laki eruption in Iceland during 1783 was likely responsible for the lowest average winter temperature in eastern United States by about 4.8°C (8.6°F). Europe also suffered from unusually frigid winter conditions. This has been attributed to a haze effect global cooling from dust and gases from the largest outpouring of basaltic lava in historic times. Large volumes of sulfur dioxide discharged into the atmosphere are considered to be the main solar radiation blocking/reflecting agent.

In 1815, Mt. Tambora, Indonesia erupted in what scientists believe was the largest eruption in 10,000 years. A result of the mass of gases and volcanic ash blasted into the atmosphere was a strong haze effect and a year without a spring and summer. Northern New England (USA) and Europe felt the global cooling most as snow fell and frost was common during June, July, and August. This affected livestock and the growth of most grains severely. Without grain, cattle could not be fed and as a result were slaughtered. The cooling also correlated with down slope advances of alpine glaciers to lower than historical measured elevations and with icebergs in Atlantic Ocean shipping lanes.

Krakatau volcano, Indonesia, erupted in 1883. It was the second largest eruption in 10,000 years. Dutch scientists calculated that 5 cubic miles (almost 21 cubic kilometers) of ash, pumice, and rock, plus huge masses of gases ejected into the atmosphere. It is likely that the haze effect from the gases and ejecta caused a global cooling during 1884 estimated at 0.5-1°C (0.9-1.8°F) during 1884 followed by unusually cool weather in the later 1880s.

In more recent times, Mt. Pinatubo in the Philippines erupted in 1991 and emitted the largest sulfur oxide cloud measured in a century. The sulfur oxide and volcanic ash circled the Earth. Over the next two years, the mean global temperature cooled by about 1°C (1.8°F). At that time scientists concerned about global warming saw the Pinatubo eruption as a modern cooling event that signaled a research path to cope with global warming.

Cooling The Earth: Can Humans Use Nature's Paths?

The concept derived from these examples and other volcanic eruptions is that if a haze effect can be created by human intervention, global cooling may be able to balance and even reverse global warming. This would require the injecting of matter into the stratosphere to block and reflect solar radiation.

The global cooling caused by volcanic eruptions are short-term events whereas scientists are seeking a long-term, intervention to stabilize global warming in conjunction with global CO_2 emissions reduction. The ultimate aim would be for global cooling to reverse the warming trend and bring the Earth temperature to what it was before the global warming trend was recognized as a threat to ecosystems and populations. This would be a stabilization temperature that would be maintained without continued anthropogenic cooling.

It is interesting to note that the remediation of sulfur dioxide (SO_2) pollution began in the United States and Europe to halt the environmental damage to vegetation and aqueous ecosystems caused by acid rain. This was just before SO_2 emissions peaked in these regions in about 1980. From the late 1970s to 2000, the SO_2 emissions from the United States declined from 31,000 thousand short tons to 17,866 thousand short tons. In Europe, the SO_2 emissions fell in one decade from 61,400 thousand short tons in 1990 to 35,949 thousand short tons in 2000. During the period 1990 to 2000, SO_2 emissions from China increased from 23,375 thousand short tons to 34,204 thousand short tons. Indian SO_2 emissions follow the same trend. These figures are representative of the drop in SO_2 emissions from developed countries and the increase in SO_2 emissions from developing countries. The decrease in the haze effect in regions of the United States, Canada, and Europe corresponds with greater physical and environmental impacts of global warming.

Asian Haze and Its Environmental Effects

China, India and other Asian nations have a daunting haze effect problem that is not being seriously addressed by nations with developing economies. In a previous chapter we discussed smog, a weather-topography localized ground-hugging haze of a mix of ozone, particles, and toxic chemicals that originate from automobile exhaust and industrial emissions. Smog lasts a short time. Nonetheless, it has caused bronchial and cardiovascular sickness and death to people susceptible to such diseases who are exposed to the deadly mix that engulfs a population center.

There is a similar but more deadly and long-term threat to populations and ecosystems in an area extending from Asia to the Persian Gulf. At the end of 2008, the United Nations Environmental Programme (UNEP) released a report that described a haze of "brown clouds of pollution" comprised of soot, particles, and chemicals up to 3 kilometers (1.6 miles) thick hanging over Asia.[55] The pollutants originate mainly from vehicle exhaust, factory emissions, and indoor cooking. The UNEP estimates that as many as 340,000 people die each year in China and India from cardiovascular, respiratory and other diseases linked to exposure to the pollutants. This represents about 0.01% of the population in these two most populated countries. It would seem to be less of a priority in these countries to keep citizens healthy than to advance industrial development rapidly. The report explains that the pollutants that comprise the brown cloud haze damage crops, change weather patterns, and are amplify the effects of global warming in the Himalayas and other glacial mountainous regions. Glacial ice reflects the sun's heat back into the atmosphere thus working against global warming. However, soot and other dark particles deposited on glaciers absorb heat and enhance melting. This is a threat to future water resources and crop yields across Asia. Thus, on the one hand, the haze provided by the "brown clouds of pollution" reduces global warming but on the other hand has negative impacts such as causing health problems and increased glacial melt. The report states that the clouds would disperse in a matter of weeks if the sources ceased to pollute. However, the emerging economies in Asia have not significantly ceased or at least slowed activities that are responsible for the pollution even at the expense of lives and future water resources and crop yields that will be necessary to sustain growing populations in Asia. This indicates that developing nations in Asia and else where have not found a less carbon intensive path to industrialization and are likely waiting for the developed nations to provide them with one or more such paths.

Planning For Haze Creation

There is a fundamental question that has to be asked and answered before embarking on a program of bringing about a haze event. What effect will the possible haze-creating matter (e.g., aerosols such as sulfur dioxide or sea water mist, and volcanic dust (ash) have on the environment when they interact long-term with the earth atmosphere, hydrosphere, and biosphere? For example, sulfur dioxide discharged into the atmosphere reacts with solar radiation to create sulfur trioxide. The sulfur trioxide reacts with moisture (water) in the atmosphere to form sulfuric acid, the main component of environment-damaging acid rain. This would rule out the use of sulfur dioxide as a long-term haze event-creating agent.

As with the extraction of CO_2 from the atmosphere, there are questions to be resolved concerning a use of the haze effect. First, scientists must determine what haze-creating matter or matter combination can be discharged into the atmosphere without harming Earth environments. A second question is how much of it should be released. Third, scientists have to determine when, where, and over what time span haze-creating matter could be released safely into the atmosphere.

Increase or Create Cloud Cover

Meteorologists noted that low altitude stratocumulus clouds (base generally at 1000-7000 ft or 300-2400 m) cover about a quarter of the world's oceans and that they reflect solar radiation. It is possible that their reflecting capability can be increased several percent by spraying a fine mist of seawater into the atmosphere to increase the area the clouds shield from solar radiation and by creating clouds that do the same. This is proposed as a way to reduce or perhaps reverse global warming by increasing blockage of solar radiation by reflecting it back into the atmosphere.[56] Some researchers estimate that this might be done by a fleet of 1900 unmanned wind-powered ships releasing sea spray mist into the atmosphere through tall funnels as they crisscross the world's oceans perpendicular to dominant wind patterns. The cost is estimated at $4.7 million per ship. This translates to a total cost of $9 billion dollars with a 25-year launch date. Nonetheless, $9 billion is a small price to pay for helping to slow and stem global warming compared to a proposed $250 billion annual cost during a 10 year period to cut the emissions of CO_2 and other greenhouse gases at the sources (e.g., industrial, transportation).[56] This out-of-the-box concept is being studied by researchers in the U.S. supported by a grant from the Carnegie Institution, and in the U.K. with a grant from Meriaura, a Finnish shipping company. The aim of any of the proposed climate engineering or solar radiation management projects is to block a few percent of the sun's radiation from reaching the earth surface and thus reduce global warming. Cost/benefit calculations have been made for several of schemes for solar radiation management and the greatest benefit of 5000 to 1 lies with using the unmanned wind-powered ships to release sea mist spray into the atmosphere.[56] Any program to control global warming would have as an ultimate goal a reduction of CO_2 concentration in the atmosphere to what some consider are tolerable stabilization levels (e.g., 325 ppm CO_2).

There are questions to be answered before a pilot study can be done to evaluate the effectiveness of the technique to create clouds or to expand the area coverage of existing clouds. What altitude should the seawater spray reach? How much spray is necessary to do the job in terms of increasing cloud cover size and hence their blocking and reflectivity efficiency? How often should the misting take place to maintain or expand an existing cloud cover? Meteorologists and atmospheric scientists are working in research teams to find answers to these and other pertinent questions.

AFTERWORD

Crises bring out the adaptability and creativity of people whether in local, national, or regional settings. Now these human qualities have to contend with the global problems that will affect all societies directly or indirectly, but affect them nonetheless. First is to contend with the growing global population in all phases of existence whether providing sustenance, clean and safe environments, and socio-economic necessities of a good quality of life. Second is to contend with global warming/climate change so as to slow the rate of warming and ultimately to stop it or ideally reverse it to a stable level such as what temperatures were during the 1950s. Human innovation can attain these ends and solve other like problems only if all nations agree to work together.

Reduction of the rate of global warming and keeping it below a tipping point that is theorized to be 2°C is one major problem nations worldwide have to attack together. Another major problem many nations have to undertake is a braking of population growth to hold the global population steady at a sustainable level. As noted earlier in the book, some believe that this may be 7-8 billion people and others 9 billion people. Solutions to these problems form the foundation for preserving the Earth as a welcoming habitat for future populations. As these conditions are being met, we can continue to work towards providing sustenance to all in minimally polluted environments in stable, lawful, and socially responsible communities. Such societies will assure sustenance for citizens, respect human and property rights, make healthcare and education readily available to all, and present an economic environment that will attract investment and hence increase employment opportunities. Freedom from want, freedom from fear, freedom of religious, freedom of speech are cornerstones of responsible long-enduring societies for all citizens, their children, their grandchildren, their great-grandchildren and later generations.

CHAPTER 8. IN-THE-BOX BRAINWORK - THE COPENHAGEN CLIMATE SUMMIT DECEMBER 2009: ASSESSMENT

KYOTO TREATY ASSESSMENT

The 1997 Kyoto Treaty called for 37 industrialized nations to work together to slow global warming and stabilize it below the 2°C that scientists set as a critical temperature increase. It required that industrialized nations reduce their carbon emissions (CO_2 and other greenhouse gases). The reduction for the European Union was set an 8% less than 1990 levels and 7% less for the United States, a non-signatory to the Treaty. The goal of the Treaty was a 52% overall global reduction in carbon emissions by 2012. The Kyoto Treaty exempted developing nations from such restrictions. The result was that from 1994-2004 the emerging economies increased their emissions. For example, China's output of CO_2 increased 4% annually during the 1994-2004 period.

By 2007, the industrialized nations' emissions declined by an average of 4% and it appeared that to some degree the Treaty was taking hold. However, by 2008/2009, 18 of the 37 industrialized countries emitted more greenhouse gases than in 1990, some substantially more. The United States emissions were 17% higher than in 1990 while other nations such as Spain and Australia had higher percent increases. China, exempted from the Treaty as a developing nation, continued to increase its amounts of carbon emissions and today spews out more than the United States to become the leading global emitter of greenhouse gases. This is a negative result of a two-tiered system that did not work: regulated nations and exempt nations. The Kyoto Treaty terminates in 2012. Clearly, a new international agreement had to be drafted to include all nations in workable greenhouse gases reduction efforts even if it means slowing economic development. Any new treaty has to be ratified by governments worldwide regardless of politics and economic status if slowing global warming is the international goal.

COPENHAGEN CLIMATE SUMMIT

To this end, the United Nations sponsored the Copenhagen Climate Summit during two weeks of December 2009 with the overriding theme of unifying actions by all nations to combat global warming. This is the follow up to the Kyoto Treaty for which targeted national and international goals designed to reduce the mass of CO_2 in the atmosphere for the most part were not met. The Summit focus was to evoke a global consensus on an immediate response to attack the problem of global warming and climate change.[57] Consensus can be achieved only if all nations sacrifice short-term economic national goals for future long-term substantial socio-economic and political gains. It is in the future that countries will most need the gains to cope with the demands of expanding populations and development needs. Ideally, any consensus reached cannot be one that is morally or ethically driven alone but one that is sealed in a legally binding treaty for developing as well as developed nations. Anything less will not slow the rate of global warming and the consequences it portends for humankind. Anything less would be detrimental to future global generations particularly in nations where emerging economies and populations are growing concurrently.

CO_2 Facts

The fact is that CO_2 is increasing in the atmosphere by 2.3 ppm annually or double that in the 1970s. This is due in part to a continued, albeit reduced input by developed nations, in part to the race to increased industrialization by developing countries, and in part to the lessening capacity of CO_2 sinks. Scientists calculate that a CO_2 atmospheric concentration of between 450 and 500 ppm would drive the global warming increase beyond the 2°C limit. If emissions are not reduced, this concentration will be met in two generations. A target date to stabilize global warming is estimated to fall between 2035 and 2050.

The Carbon Intensity Machiavellian Maneuver

Previous to the beginning of the December 2009 Copenhagen Climate Summit, Premier Wen Jiabao of the People's Republic of China announced that his country would work towards curbing greenhouse gases emissions. China would cut energy use by 20% and set a carbon intensity target goal of 40-45% reduction by 2020.[58]

India set its carbon intensity goal at 20-25% by 2020. The decline of energy use is certainly a positive action. However, couching meaningful reduction of CO_2 as a carbon intensity target goal percent is somewhat deceptive as a measure of absolute numerical tonnage reduction of CO_2 emissions.

Carbon intensity can be defined as the ratio of carbon emissions to economic activity. It has also been defined as a measurement of carbon emissions versus primary energy use, or as energy use per unit of gross domestic product. Thus, carbon intensity measurements are a reflection of the efficiency of an economy with respect to carbon emissions. However, a carbon intensity reduction does not mean an actual reduction in carbon emissions. For example, during the decade 1990-2000, the carbon intensity of the United States economy declined by

17% but total missions increased by 14% due in grand part to a 39% increase in economic activity.[59]

The use of carbon intensity has been called a bookkeeping maneuver that self-serving governments can use to suggest that they are maximizing efforts to slow global warming. This really means that as energy efficiency in various economic ventures improves, carbon intensity lessens, profits increase for investors, and government coffers benefit from taxes on the added profits. If production increases to maintain a high rate of economic growth, there will not be a large decline in the actual mass of carbon emissions. Thus, increasingly damaging levels of CO_2 and other greenhouse gases continue to invade the atmosphere to fuel global warming and drive the ensuing changes in climate. The People's Republic of China goal of reducing carbon intensity by 40-45% by 2020 as the country maintains a strong rate of economic growth (better than 9% during 2008; 8.7% for 2009 with a 10.7% fourth quarter). This means that absolute tonnage of emissions will continue to increase but at a somewhat slower rate. The same is true for India and other nations that have not already taken severe steps to reduce the masses of carbon they emit to the atmosphere.

THE COPENHAGEN ACCORD

The product of the Summit was the Copenhagen Accord, a short document that is an agreement by participating nations to draft a program in some detail during 2010 that would address several themes debated at the Summit. It put forth the major points that were agreed to as well as disagreements that had to be resolved between the more than 190 nations represented in Copenhagen. For example, the plan would establish carbon emission targets for developed and developing nations that would not disrupt a fair rate of sustainable development.

The question here is what percent of economic growth is rational to achieve this end. Brazil has pledged to an economic growth rate of a reasonable 3.5%. If China seeks to maintain its 2009 economic growth rate during 2010 and subsequent years, this would surely derail the international effort to stem global warming. This, in spite of the fact that it is in China's near- and distant-future benefit to accede to a lower economic growth rate, perhaps 5-6%. The plan is to lay out norms to help poorer countries to adapt to climate change. The following pages describe several of the principal points.

The Paths to Reach Greenhouse Gases Reduction Goals

Only a multi-pronged approach can achieve the goal of curbing the immediate and long-term effects of global warming by reducing CO_2 emissions before reaching the critical 2°C increase in warming. The plan is to aim for the lowest emissions that will be necessary to allow a sensible rate of sustainable development. The prongs focus on mental/psychological, economic, social, technological, and physical parameters.

Conceptual Paths

First, governments worldwide have to realize that keeping the global warming tipping point temperature below 2°C, perhaps at 1.6°C requires a long-term, continuing effort. They agree on this. Second, they must recognize that without a legally binding treaty, the likelihood of achieving the goal is in question. This is a major problem because representatives of China and India indicated that their nations would not sign a treaty with this stipulation. Many other developing nations agree with this stance. Without a concession on this point, the probability of a functional treaty will go the way of the Kyoto Treaty, generally unsuccessful except for bringing to light the threats that global warming and ensuing climate changes pose to all nations.

Third, the industrialized nations have to accept the responsibility for adequate, predictable, and sustainable financing, technology assistance, and capacity building for adaptation to the immediate and future effects of global warming and the ensuing climate change. This is sensible particularly when most of these nations do not have excess foreign reserves (with respect to foreign debt) to use to help themselves. However, it is also sensible to expect that high-income developing countries with large foreign reserves and low external debt contribute significantly to the funding for adaptation to and mitigation of global warming. For example, China has a foreign exchanges reserve of 2,399,000 million of US dollars (December 2009) versus an external debt 363,000 million of US dollars (December 2007) for a ratio of almost 25:1. In comparison, the United States has a foreign exchange reserve of 83,375 million of US dollars (July 2009) and a foreign debt of 13,773,000 US dollars (June 2009) for a ratio of 1:165. Similarly, the United Kingdom has a foreign exchange reserve to external debt ratio of 1:151. A level of new and additional money was proffered and seems workable. The amounts proposed are $30 billion for 2010-2012 for adaptation and mitigation in the most vulnerable countries (especially in sub-Saharan Africa) and small island states. Another $100 billion a year is intended to be granted by 2020 for least developed and low-income developing countries. It would seem that China should be making significant a contribution to a global warming adaptation and mitigation fund commensurate with its global economic position cited above. Similarly positioned countries should do the same. This proposed contribution from high-income developing countries is likely to be contested in future negotiations. In no case should any funding be tied to political agendas or influence on access to natural resources. A problem to be resolved is the entity controlling the distribution and monitoring of the use of fund to verify that they are used for the purposes for which they are granted. A Copenhagen Green Climate Fund was proposed with administration by the United Nations. However, donor countries know the grave non-transparency and corruption/bribery problems that colored the United Nations administration and handling of Iraqi oil sales and funds a few years ago. They want to avoid this. Thus, an independent, overview control committee of experts in various aspects of adaptation and financing, thoroughly vetted by the donor states (outside United Nations influence), could be formed. This committee perhaps with input by a World Bank-IMF unit would likely be the administrative answer to obviate any mishandling or misappropriation of funds.

Fourth, all nations have to accept the fact that only by reducing emissions of CO_2 and other greenhouse gases from their particular national sources can they contribute to the curbing of global warming. Initially, the mitigation should be voluntary for the least developed and small developing states. In addition, developed countries should also provide incentives (monetary and market access) to low-income developing countries with low carbon emitting economies to stay with these. For other developing countries that have not found a less carbon intensive path to industrialization, developed countries should provide them with technology and capacity to follow such paths. Fifth, emissions mitigation by all countries should be subject to international measurement, reporting, and verification of the progress of nationally appropriate mitigation actions. This is likely acceptable on the basis of verifiable reports at two year intervals.

Sixth, forests, major sinks for CO_2, should be preserved and reforestation of stands that have been ravaged is a top priority for all nations. This seems to be an attainable goal if the industrialized nations pay to preserve forests where they are at risk and for reforestation where deemed necessary. Otherwise, least developed and developing countries might allow forests to be cut down for wood, farmland, and grazing land.

Importantly, all nations have to agree on carbon markets, the so-called cap and trade procedures and their monitoring. Cap and trade is designed to prevent emissions that will increase the masses of CO_2 and other greenhouse gases in the atmosphere but will do little to achieve the aim of the Copenhagen Accord: to reduce the masses of these gases in the atmosphere. Cap and trade is sort of a half-baked contribution to curb global warming. Only reduction in actual tonnages of carbon emissions into the atmosphere with respect to contemporary levels can work to counter global warming.

Practical Paths

The goals proposed during the Copenhagen Climate Summit are far less than scientists calculated are necessary to stabilize the atmospheric CO_2 content in a fixed time frame (2030/2050) in order to keep warming below 2°C.[60] Nevertheless, we have to consider several ways that countries can activate actionable efforts to meet CO_2 reduction targets set by law or treaty or pledged under binding agreements by responsible governments. Most of these will require international technical and economic assistance. Carbon dioxide emission reduction methods financing can be used to various ends. Initially, these should be for replacing old inefficient (dirty) power plants with efficient modern ones built with CO_2 capture/conversion emissions control systems, and for replacing energy inefficient transportation modes (vehicles, aircraft, ships) with efficient ones. These methods can be complemented by implementing energy conserving building codes (at no cost), and by replacing traditional coal- or charcoal-based cooking and heating stoves with efficient ones or stoves based on renewable energy where applicable (e.g., solar cookers being used in Darfur, Sudan). In addition, funding for research in clean, renewable energy technologies, and tax or other incentives to stimulate energy efficiency will help reduce atmospheric CO_2 contents. The community of nations hopes to bring down carbon emissions to the atmosphere fast enough to curb the effects of global warming and of climate change by making these and other energy cutting changes an immediate priority.

We will know by the end of 2010 or more likely during 2011 whether there will be a legally binding Copenhagen Climate Treaty that will definitely cut carbon emissions to lower levels than they were in 2009 and proceed to further reduce them at the least to lower than 1990 tonnages. This can be done with reasonable economic sacrifice by all but sacrifice that allows for fair national sustainable development. Concurrently, the community of nations can continue the efforts towards providing sustenance, and social, economic, and political security and stability for a growing global population in environments safe from harmful pollution.

NOTE: JANUARY 2010

As United Nations directed negotiations proceed on the several major points in the Copenhagen Accord, most nations believe that there is little chance of meaningful agreements. Because the main outcome of the Copenhagen Climate Summit was to urge deeper cuts in CO_2 and other greenhouse gases but not demand these cuts by binding legal agreement, the drive to slow global warming through the efforts of the Summit is likely a lost cause, another Kyoto Treaty. In anticipation of the lack of success through the United Nations sponsorship, the European Union intends to pursue a new approach through the Group of 20 (G-20) by exploring other meaningful options that will result in an international rapprochement for bringing global warming under control. The G-20 is comprised of 19 countries and European member nations that are responsible for 85-90% of the world's gross international product (the market value of all final goods and services produced in the world in a given year). The G-20 members are also responsible for 80% of global trade, and have 2/3 of the earth's population. Only a focus on economic issues that will arise from global warming, country by country, can bring nations with developing and emerging economies out of a nationalistic tunnel vision into an international futuristic mode. To envision the future, really understand what will be, and present this vision to the world's people in intelligible media form, should bring internal and international pressure on governments to join a workable global warming treaty. The G-20 has the economic power to make things happen, such as a legally binding treaty to greatly reduce globally the driving forces of global warming, CO_2 and other greenhouse gases emissions.

EPILOGUE

The reality checks in our contemporary environmentally troubled world, give we the people, and the governments that represent us, action/no-action choices. We can elect to live in a "Do Nothing Future" and hope that nature and good sense will solve the problems of population growth, of physical, biological, and chemical degradation in ecosystems vital to human existence, of disparate economic development, and of socio-political unrest versus social stability. This is not likely to happen. Instead global conditions will surely deteriorate into a state of heightened competition and perhaps chaos among people, and conflicts between nations vying to survive. A second choice is to live in a "Do Little Future" with the same results as above and delayed somewhat but with problems left unresolved. Our final option and the only one acceptable if we are to maintain a good quality of life as many now know it and extend it to communities where it is lacking, is the "Do Now Future". We must meet the challenges facing global populations head-on, together as a worldwide citizenry, and wholeheartedly contribute our collective knowledge, labor, and treasure to solve natural and anthropogenic problems that threaten lives be they local, regional, national or global. Such actions must start now, when it is still possible to stunt the advances of processes that threaten humankind. Obviously this is preferred to grappling with threatening situations in the future when it might be too late to effectively bring about necessary changes to help populations in distress.

To be sure, there are multi-pathways humans trod that can take us further towards environmental degradation and global disasters if they converge at one of several at risk regions on our planet. There is one pathway carrying expanding populations, pollution of our ecosystems, and the inability to provide sufficient and safe water and food for all people. Another pathway bears global warming and ensuing climate change and the ills they impose on societies. Contagious diseases, some without therapies, and chronic illnesses of various origins move on a third pathway. Still a fourth conveys a global condition fraught with repression of citizenry, religious and ethnic zealotry, and unyielding dogma. These and other pathways threaten the physical, chemical and biological fabrics that together support ecosystems and sustain life. Elimination of the threats to populations and their environments will prevent collapses and crashes of our ecosystems. This will alleviate the physical, biological and psychological stresses on ecosystems and people worldwide.

Yet, what is being done and what will be done to regain the balance between humans and their environments remains a Gordian knot with a few strands freed and others loosened but not untied. The knot can be undone only by cooperation between developed, developing, and less developed nations alike. Collaborative efforts alone can move humankind back from tipping points of disaster that threaten the continued existence of 100s of millions if not one or two billion or more humans on our Earth now and in the future.

The elements leading to disasters can be brought to heel but only by total collaboration. This can only be guaranteed by treaties with enforcement clauses to call to task those nations that try to obviate a treaty though subterfuge. Any method used to explain why a country or group of countries should not adhere to conditions that are designed to preserve long-term sustainability of humans and their environments is not defensible. As noted earlier in the book, China, with its population of 1.35 billion inhabitants (2010) emits more CO_2 from its coal-fired power plants and other industries that combust fuels than any other country in the world. China defended its position in the past by stating that its per capita amounts of CO_2 emissions are low...clearly this is subterfuge because in our world of global warming and climate change, per capita emission weighs nothing against the total mass of CO_2 emitted. Not slowing the rate of global warming and the consequent climate change means that if there is no collaborative global action, China will suffer longer and more closely spaced drought, more killer floods, more air and water pollution contamination, and will continue to lose productive farmland as it has been losing for several years running. Other developing nations will suffer from similar climate change problems unless they are willing to forego a high rate of economic development for a moderate rate, so as to assure their socio-political and economic futures.

The rationale of many developing nations that industrial nations achieved their development aims by polluting the environment. They argue that developing and less developed nations should be given latitude to do the same. This attitude is ill conceived given the state of environments worldwide. In the end it is self-defeating as exemplified in the pollution and farmland loss problems experienced by China as the government pushes to maintain or increase its 2009 8.7% rate of economic growth. This is in light of the realism that reasonable growth can continue at a subdued rate, perhaps 6%, or 4%, but growth nonetheless. This can help to relieve stress on ecosystems that sustain humans and other life forms/natural resources. Developed (industrial) nations have to keep reducing the inputs of damaging pollutants that harm environments beyond the extraordinary reductions many have already made (e.g., SO_2, Hg, other potentially toxic metals) and continue to make. China's weak responses to environmental intrusions complemented by those of other developing nations (e.g., India) and less developed nations now negatively impact the quality of life conditions mainly for their citizens. However, they also affect citizens worldwide. This can only increase in the future unless all nations come together to confront known global situations that menace global populations.

There are precedents for the nations of the world to unite and collaborate to solve serious global problems that affect global populations, some to a great degree, others to a lesser degree. One, as discussed in Chapter 3, is the positive effort by most nations to gradually but surely bring down fertility rates and hence slow the rate of population increase until the replacement level is reached and global population is stabilized. This was in response to projections of future problems individual nations and the global community would encounter in order to sustain growing populations. Another precedent is in 191 nations joining and adhering to a treaty to reduce and then eliminate the causes of the thinning and breaching of the Earth's protective ozone layer. This effort has been successful and the ozone layer appears to be healing. The factor that is holding back unified action to first slow the rate of global warming, stop it, and ultimately reverse is national interests. These are mainly economic interests as they influence a nation's socio-political stability at this moment in time. Many government leaders fail to consider the future expense to their citizens' children, grand children, and generations as they fail to confront global warming and other worldwide problems.

One can justifiably question, where is the balance between a good life and development realities? What will happen to urban and rural populations? Is there a future sustainability problem? Is rapid development worth what could be grand, frightening sacrifices of populations, or is slowing development and caring for people and their environments a better plan for the future?

You know the answers to these questions. However, the main question remains: How do we get leaders or leadership with short-term national goals to react to the questions above with answers that will assure long-term, future benefits for their countries?

SELECTED REFERENCES

[1] Carson, R.L., 1962. Silent Spring. Houghton Mifflin, Boston, MA, 368 p.

[2] Ehrlich, P.R., 1968. The Population Bomb. Ballentine Books, New York, 223 p.

[3] Ehrlich, P.R. and Ehrlich, A.H., 1990. The Population Explosion. Simon and Schuster, New York, 320 p.

[4] Population Reference Bureau, 2009. 2009 World Population Data Sheet. 16 p., Washington, D.C. Can be viewed on line at http://www.prf.org

[5] United Nations, 1998. World population Prospects: The 1998 Revision. Source: http://www.uneca.org/.../programme_overview/population/ fertility/fertility...world.htm

[6] World Bank, 2004. World Population Growth. Can be viewed on line at http://www.worldbank.org/depweb/english/beyond/global/beg-en.html#toc

[7] United Nations Population Divison, 2009. World Population Prospects. The 2008 Revision, medium variant. Can be viewed on line at http://www.un.org/esa/population/wpp2008_highlights.pdf

[8] Population Reference Bureau, 2008. Nutrition. World Population Highlights, 2008. Population Bulletin, vol. 63, no. 3, p. 7, Washington, D.C. Can be viewed online at http://www.prb.org/pdf08/63.3highlights.pdf

[9] U.S. Census Bureau, 2009. International Data Base 2008. For Figures 1-10. Can be viewed on line at http://www.census.gov/ipc/www/idb/country.php

[10] Population Reference Bureau, 2008. World Population by Region or Nation. Population Bulletin. Vol. 63, No. 3, p. 3. Prepared from Data in United Nations Population Division, 2007. World Population Prospects. The 2007 Revision (medium scenario). Can be viewed on line at http://www.prb.org/pdf08/63.3highlights.pdf

[11] Population Reference Bureau, 2009. Numbers of Elderly Relative to Young in More Developed and Less Developed Countries. Population Bulletin, Vol. 64, No. 3, p.3. Prepared from data in United Nations Population Division, 2008. World Population Prospects. The 2008 Revision (medium scenario). Can be viewed on line at http://www.prb.org/Publications/PopulationBulletin/2009/ worldpopulationhighlights2009.aspx

[12] World Resources Institute, 2000. World Resources 2000-2001. People and cosystems: The Fraying Web of Life. Collaborative Effort of United Nations Development Programme, United Nations Environmental Programme, World Bank, World Resources Institute, Washington, D.C., 389 p. Can be viewed on line at http://www.wri.org/wr2000/

[13] Pimentel, D., Cooperstein, S., Randell, H., Filberto, D., Sorrentino, S., Kaye, B., Nicklin, C., Yagi, J., Brian, J. and O'Hern, J., 2007. Ecology of increasing diseases: Population growth and environmental degradation. Human Ecology (Springer), 35: 653-668.

[14] Food and Agriculture Organization of the United Nations, 2009. The State of Food Insecurity in the World 2008. Updated via FAO news release, 19 June 2009.

[15] World Health Organization, 2008-2009. Cholera in Zimbabwe. Updates. Source: http://www.who.int

[16] Harden, B., 2009. Japan works hard to help immigrants find jobs. Washington Post, Jan. 23, 2009. With contribution from Yamamoto, A.

[17] National Corn Growers Association, 2007. Corn Growers Reducing rosion. Source: californiafarmer.com/story. aspx?s=130488c=8 See also USEPA Agricultural Center Soil Preparation report. Source: www.epa.goc/agriculture/ag101/cropsoil.hmtl

[18] Worm, B., Barbier, E.B.,Beaumont, N., Duffy, J.E., Folke,C., Halpern, B.S., Jackson, J.B.C., Latze, H.K., Micheli, F., Palumbi, S.R., Sala, E., Selkoe, K.A., Stachowicz, J.J. and Watson, R., 2006. Impacts of Biodiversity Loss on Ocean Ecosystem Services. Science, 314: 787-790.

[19] World Resources Institute, 2008. World Resources 2008 - Root of Resilience - Growing the Wealth of the Poor. World Resources Institute in collaboration with the United nations Development Programme, the United Nations Environmental Programme, and the World Bank, Washington, D.C., p. 210-213. Can be viewed on line at http://www.wri.org/publication/world-resources-2008-roots-of-resilience

[20] International Panel on Climate Change, 2007. Climate Change 2007 (as 4 part report). Part 1. The Physical

Science Basis (February 2007); Part 2. Impacts, Adaptation and Vulnerability (April 2007). Part 3. Mitigation of Climate Change (May 2007). Part 4. Synthesis Report (November 2007). World Meteorological Organization and United Nations Environmental Programme. Source: http://ipcc.ch/pdf/assessmentreport/ar4/syr/ar4_syr_spm.pdf

[21] Pigeot, J., 2000. Le Figaro, June 20, p. 14.

[22] Cyna, B., Chagneau, G., Bablon, G. and Tanghe, N., 2002. Two years of nanofiltration at the Méry-sur-Oise plant, France. Desalination, 147: 69-75.

[23] Blachman, A., 2008. How to make water from thin air. Source: cleantechnica.com/2008/11/05/how-to-make-water-from-thin-air

[24] Waldoka, E.Z., 2009. In world first, Israeli know-how helps Indian village get water from air. Jerusalem Post. Source: http://www.jpost.com/servelet/Satellite?cid=1233304849721&pagenames=JPost%2FJPArticle%2FShowfull

[25] Food and Agricultural Organization of the United Nations, 2008. The State of Food Insecurity in the World 2008. Food and Agricultural Organization, Rome, 56 p. Can be viewed on line at http://www.fao.org/decrep/011/10291e00.pdf

[26] Grimaraes, E., Ruane, J., Scherf, B., Sonnino, A. and Dargie, J. (eds.), 2007. Marker-Assisted Selection. FAO of the UN, Rome, 441 p. Can be viewed on line at http://www.fao.org/docrep/010/a1120e00.htm

[27] SANET, 2006. GM ban overdue. Source: http://lists.ifas.ufl. edu/egi-bon/wa.exe?A2=ind0602&L=sanet-mg&P=3448

[28] Prescott, V.E.,Campbell, P.M., Moore, A., Mattes, J., Rothenberg, M.E., Foster, P.S., Hihhins, T.J.V. and Hogan, S.P., 2005. Transgenic expression of bean a-amatase inhibitor in peas results in altered structure and immunogenicity. Jour. Agriculture and Food Chemistry, 53: 9023-9030.

[29] Food and Agricultural Organization of the United Nations, 2008. FAO Statistical Yearbook 2008. FAO, Rome, Italy. Can be viewed on line at faostat/fao.org/

[30] UNEP/GRID-ARENDAL, 1997. Atlas of Desertification in the World, 2nd Ed. United Nations Environmental Programme, Arnold Publishers, London, 182 p.

[31] Oldeman, L.R., 1994. The Global Extent of Soil Degradation. In Soil Resilience and Sustainable Land Use. CAB International, Oxon, U.K., p. 115.

[32] Delgado, C.L., Wada, N., Rosegrant, M.W., Meijer, S. and Ahmed, M., 2003. Fish to 2020: Supply and Demand in Changing Global Markets. International Food Policy Research Institute, Washington, D.C. and World Fish Center, Malaysia, 232 p. Can be viewed on line at http://www.ifpri/ publication/fish-2020 http://www.ideas.repec.org/b/wfi/ wfbook/15796.html

[33] Worm, B., Hilborn, R., Baum, J.K., Branch, T.A., Collie, J.S., Costello, C., Fogarty, M.J., Fulton, E.A., Hutchings, J.A., Jennings, S., Jensen, O.P., Lotze, H.K., Mace, P.M., McClanahan, T.R., Minto, C., Palumbi, S.R., Parma, A.M.,

Ricard, D., Rosenberg, A.A., Watson, R. and Zeller, D., 2009. Rebuilding Global Fisheries. Science, 325: 578-585.

[34] World Health Organization, 2009. WHO Report 2009: Tuberculosis Control - Epidemiology, Strategy, Financing. WHO, Geneva, Switzerland, 303 p. Can be viewed on line athttp://www.who.int/tb/publications/global_report/2009/pdf/full_report.pdf

[35] Snow, R.W., Guerra, C.A., Noor, A.M., Myint, H.Y. and Hay, S.I., 2005. The global distribution of clinical episodes *Plasmodium falciparum* malaria. Nature, 4334: 214-217.

[36] World Health Organization, 2009. AIDs epidemic update: December, 2007. WHO, Genva, Switzerland, 50 pp. Can be viewed online at http://www.who.int/hiv/

[37] Gaffin, S., Hachadoorian, L. and Engelman, R., 2006. Mapping the future of world population 2006. Collaboration between the Center for Climate Systems Research at Columbia Univ. and Populaion Action International. Population Action International, Washington, D.C. Can be viewed on line at http://www.earth.columbia.edu/news/2006/story07-11-06.php

[38] DeCarbonnel, E., 2009. Catastrophic fall in 2009 global food production. Source: http://www.information clearinghouse.info/article21955.htm

[39] Lau Kin Chi, 2009. Food security and sustainable livelihood in China. Taken from the Chinese Ministry of Land and Resources press release 16 April 2008. Can be viewed on line at http://www.landaction.org/spip/php?article442

[40] Montreal Protocol (1987). Full text can be viewed on line Source: unstats.un.org/pop/Documents/Doc0007.htm

[41] Bodman, S.W., 2008. Remarks. At 2008 Energy Conference: International Energy Outlook 2008. Energy Information Administration of the U.S. Department of Energy, Washington, D.C., April.

[42] United Nations Environmental Program, 2001. Climate Change 2001: Working Group 1: The Scientific Bassi. Technical Summary of the IPCC Third Assessment Rrport. Can be viewed on line at http://www. grida. no/publications/other/ipcc_tar/?src=/climate/ipcc_tar/wg1/017.htm

[43] Wynn, G., 2009. Two meter sea level rise unstoppable: experts. Reuters, September 30, 2009. Can be viewed on line at http://www.reuters.com/articlePrint?articleId=USTRE5 8S4L420090930

[44] Costello, A., Abbas, M., Allen, A., Ball, S., Bell, S., Bellamy, R., Friel, S., Grace, N., Johnson, A., Kett, M., Lee, M., Levy, C., Maslin, M., McCoy, D., McGuire, B., Montgomery, H., Napier, D., Pagel, C. Puppim de Oliveira, J.A., Redclift, N., Rees, H., Rogger, D., Scott, J., Stephenson, J., Twigg, J., Wolf, J., and Patterson, C., 2009. Managing the Health Effects of Climate Change. The Lancet, 373: 1693-1733.

[45] Lex, N., 2009. Canadian scientists breeding cows that burp less. Reuters report, June, 2009. Can be viewed on line at http://www.canadian.com/entertainment/books/Canadian+scientists+breeding+cows+that+burp+less/

[46] Nkrumah, J.D., Okine, E.K., Mathison, G.W., Schmidt, K., Li, C., Basarab, J.A., Price, M.A., Wang, Z. and Moore, S.S., 2006. Relationship of feedlot feed efficiency, performance, and feeding behavior with metabolic rate, methane production, and energy partitioning in beef cattle. Jour. Animal Science, 84: 145-153.

[47] Donoghue, E.R., Nelson, M., Rudis, G., Sabogal, R.I., Watson, S.T., Huhn, G. and Luker, G., 2003. Heat-related deaths - Chicago, Illinois 1996-2001, and United States. CDC MMWR Weekly, July 4. Can be viewed on line at http://www.cdc.gov/mmwr/preview/mmwrhtml/mm5226a2.htm

[48] De Bono, A., Peduzzi, P., Kluser, S. and Giuliani, G., 2004. Impacts of Summer 2003 Heat Wave in Europe. Environment Alert Bulletin 2, 4 p. United Nations Environmental Programme, New York. Can be viewed on line at http://www.grid.unep.ch/product/publication/download/ew_heat_wave.en.pdf

[49] Blier, W., 2009. The record breaking Central California Heat Wave of July 2006. NOAA Report 5A.8, 12 p. Can be viewed on line at http://www.ams.confex.com/ams/pdfpapers/123744.pdf

[50] Australia National Meteorological Service, 2009. The exceptional January-February 2009 heatwave in south-eastern Australia. Special Climate Statement 17, 11 p., Melbourne. Can be viewed on line at http://www.bom.gov.au/climate/current/statements/scs17d.pdf

[51] National Oceanographic and Atmospheric Administration, 2009. Global hazards and significant event in February 2009. National Climatic Data Center. Can be viewed on line athttp://www.ncdc.noaa.gov/oa/climate/research/2009/.../hazards/html

[52] Fox, M., 2009. AIDS vaccine protects people, shocks researchers. Reuters, Sept. 25, 2009, 2 p.

[53] Bremner, J., Haub, C., Lee, M., Mether, M. and Zuehlke, E., 2009. World Population Highlights - Key Findings from Population Reference Bureau World Population Data Sheet. Population Bulletin, Vol. 64, Population Reference Bureau, Washington, D.C., p. 10. Can be viewed on line at http://www.prb.org/Publications/PopulationBulletin/2009/ worldpopulationhiighlights2009.aspx

[54] International Energy Agency, 2009. Key World Energy Statistics 2009. OECD/IEA, Paris, France, 78 pp. Can be viewed on line at http://www.iea.org/publications/free_new_Desc.asp?PBS_ID

[55] Ramanathan, V.M. and team of 50 collaborators, 2008. Atmospheric Brown Clouds: Regional Assessment Report with Focus on Asia. United Nations Environmental Programme, Nairobi, Kenya, 39 pp. Can be viewed on line at http://www.unep.org/pdf/ABCSummaryFinal.pdf

[56] Webster, B. and Devlin, H., 2009. Cloud ship on course to beat climate change, says Copenhagen study. The London Times - timesonline, 7 August 2009. Can be viewed on line at http://www.timesinline.co.uk/

[57] Bickel, J.E. and Lane, L., 2009. An analysis of climate change as a response to global warming. The Copenhagen Center - consensus on Climate, Copenhapgen, Denmark, 59 p. Can be viewed on line at http://www.fixtheclimate.com/component-1/the solutions-new-research/climate-enginering

[58] Mukherjee, K., Chestney, N., Wynn, G., Doyle, A., and Fogarty, D., 2009. Analysts' View: China Announces CO_2 Intensity target for 2020. Reuters, November 20, 2009. Can be viewed on line at http://www.reuters.com/article/environmentNews/ idUSTRE5AP14D20091126

[59] Fischlowitz-Roberts, B., 2002. Carbon emissions climbing. Report from the Earth Policy Institute, Washington, D.C. Can be viewed on line at http://www.earth-policy.org/index.php?/indicators/C52/ carbon_emissions...2002

[60] Doyle, A., 2009. World carbon emissions overshoot "budget": PwC. Reuters, December 1, 2009. Can be viewed on line at http://www.reuters.com/article/domesticNews/ dINGEE5AT2N920091201

INDEX

www.ingramcontent.com/pod-product-compliance
Lightning Source LLC
Chambersburg PA
CBHW071150290526
45788CB00001BA/221